The Illusion of Superiority

by

Christopher Joseph

Copyright by Christopher Joseph © 2021

All rights reserved. No part of this book may be reproduced, stored in a retrieval system, or transmitted in any form or by any means without the prior written permission of the author, except by a reviewer who may quote brief passages in a review to be printed in a newspaper, magazine, or journal.

The author alone grants the final approval for this literary material.

First printing.

Please visit the official website of Christopher Joseph www.TheWorldHidden.com to view his photography, blog, and excerpts of his other literary works.

For my parents

and

For the Saints

Foreword
by Dr. Barrett K. Robinson, MD, MPH

This past year has seen the coronavirus knock the entire world off its axis—and the United States has been anything but exempt from the widespread loss of life, social turmoil, and economic ruin left in the pandemic's wake. Increasing partisanship, seemingly spurred by Donald Trump's rise to power and nationalist rhetoric, made heretofore unusual expressions of caustic social views both commonplace and highly incendiary. While the murder of George Floyd by Derek Chauvin illuminated the daily realities of police brutality faced by people of color, many of the conversations generated took place solely between somewhat open-minded individuals. Productive crosstalk between conservatives and liberals and between Republicans and Democrats was limited.

Many of these communities used this would-be opportunity for growth and expansion of thought as an impetus to double down and further entrench themselves in their historical philosophies. In an age where "wokeness" is often viewed as laudable or too liberal, and political correctness esteemed essential or too weak, open and thoughtful discourse about sensitive social topics with almost anyone outside of one's family structure is all too often scuppered in favor of avoiding the cardinal sin of offending the listener or intentionally being offensive.

The Illusion of Superiority is a welcome departure from this trend, as the author dives into a myriad of hot-button topics utilizing an ecclesiastical mind-set with refreshing honesty. While Christopher Joseph clarifies that he is not an authority on these issues, his profound comprehension of the subjects under discussion, clarity of thought, and panoramic approach demand the reader think long and hard about the themes upon which he touches. These are conversations that we want to have with each other. These are conversations that we need to have with each other—and these conversations would indubitably benefit from the wisdom shared within the pages of this tome.

When Christopher reached out to me about a year ago, I was somewhat taken aback. It had been almost twenty years since we last spoke. I am also a physician, and while in the midst of seeing patients one afternoon, my scheduler informed me that a physician was on hold for me. When I picked up the line, I recognized his voice instantly. Christopher had Googled me, found my work number, and decided to call me up and see how his long-lost classmate was faring. Some people use Facebook; some people use the yellow pages. Christopher was apparently more the latter than the former.

We struck up a dialogue fairly briefly because I had two patients in rooms, and they were undressed. The gowns are almost as thin as paper towels and rather drafty, so Christopher and I made plans to reconnect soon. Since then, we've done a commendable job of remaining in touch, no small task for two fathers of young kids whose hours outside of the hospital are ever at risk of being invaded by the duties and requirements of practicing medicine in the United States.

It doesn't have to be this way. That statement could apply to countless conditions, yet I'm referring to how we practice medicine in the United States. (I will spare the reader a treatise on the many flaws in our current medical model.) Still, I fondly recall the one-month-long journey out of the country I took just before starting my OB-GYN residency training. I traveled to Cuba with about a dozen other young people, most of us medical students. I was in my fourth year at Duke University, a mere three months from graduating. I'd recently earned my master's in public health in the research triangle from the "enemy" institution, just eight miles up Route 15-501 in Chapel Hill, and had grown increasingly fascinated with other countries' health care systems.

I'd spent six months interning at the Latin American Center of Perinatology, traveling throughout Uruguay and Argentina, working with birthing hospitals to improve data collection on pregnant women. In my travels, at that point visiting over thirty nations, I noticed a shortage of physicians in many of these developing nations. More often than not, when I met a doctor who was not born in the country I was visiting, they were almost always Cuban. Despite Cuba having a population of only eleven million, the eighty-third largest in the world as of 2015, this tiny nation had 37,000 Cuban medical employees working in seventy-seven countries worldwide. For almost sixty years, Cuba has exported physicians as part of numerous humanitarian missions. This arrangement benefits both the Cuban government (Cuba earns billions of dollars for providing physicians to other nations) and the Cuban doctors, who receive a much higher salary in the host nation than they would receive in Cuba.

Despite sending so many doctors abroad, health care is a fundamental right free of charge to Cuban citizens. This cash-strapped nation's ratio of one physician to every 150 citizens is second only to Qatar, the wealthiest nation in the world. It dwarfs the physician-to-population ratio by two- to threefold for countries such as the United States, Canada, Australia, and the United Kingdom. Cuban doctors often live in the neighborhoods of the people whom they serve. Yes, they have regular work hours, but patients also know that they can walk three or four blocks at 9 p.m., knock on the door of their family physician, and be seen urgently in many instances. I got the impression that this familiarity with their medical providers was highly valued and not prone to abuse. There is a genuine sense of concern for the individual by the physician, beyond the role of just being a patient, as this is their neighbor, the parent of their daughter's school chum, or their grocer. This critical emphasis on health and preventative care is one of the reasons life expectancy in Cuba is almost identical to that in the United Kingdom.

The month I spent in Cuba was anachronistic, like stepping into a time capsule buried in the 1950s—and it wasn't just the 1957 Chevy Bel Airs slowly cruising down the boulevards or the Spanish colonial architecture with vaulted arches, stucco, and courtyard fountains. It was just as much due to the amused grins of elderly men on the corner or the coquettish smiles of twenty-something-year-old maidens twirling while in the midst of a spirited danzón under a blanket of stars. Whether it was spotting children wearing cutoff jean shorts and flip-flops playing soccer in the streets or kids using a broomstick to hit a baseball in an alleyway, before sprinting around sewer grates serving as bases, strangers didn't eye you with suspicion; rather, their expressions seemed to invite you to sit down with them and sip on a Cuba libre while taking in the Malecon. Those interactions often led to casual conversations comparing and contrasting our nations, our world views, and whether I preferred the mojito or the daiquiri. My days in that beautiful nation imbued within me an appreciation for a rhythm, a warmth, a pulchritude, a unity, and even a Blackness that I found rivaled only by my time in Brazil.

It is thus evident to me that Christopher's extensive time spent in Cuba has affected him similarly, seemingly to an even greater extent. He has embraced understanding his wife's heritage, and the roots of his initial work, *The World Hidden*, can be traced directly to his exposure to this life-changing and often diametrical worldview. Only when we step outside of what we know can we step outside of who we are and truly transform our minds. I recall once hearing that "life is a book, and those who have never traveled have read only a page." *The Illusion of Superiority* clearly is authored by a man who has traveled across many cultures and has lived a very rich life at a relatively young age.

The Serenity Prayer requests God to "grant me the courage to change the things I can change, the serenity to accept the things I cannot change, and the wisdom to know the difference." Angela Davis later revised it a bit, discarding the pleas for serenity and stating that "we must change the things we cannot accept." I view *The Illusion of*

Superiority as an attempt by Christopher Joseph to do just that: to open the eyes of the reader in such a way that one comes to their own conclusion about what they can and cannot accept. The book addresses broad-ranging themes —the inevitability of death, the quest for fulfillment, the need for security, the purpose of life, laboring out of necessity, justice, the seduction of life's pleasures, health and wellness, wealth and materialism, religion, the need for love, cultural superiority, social media and its effects on children, prejudice and racism, youth and its fleeting nature, benevolence, and government structure. There is no lack of fodder here for consumption by those with an appetite for learning and challenging preconceived notions.

    I remember the first time I met Christopher. It was in our freshman year, and we were both in General Biology 101 at Morehouse College, under Dr. Clarence Clark's instruction. I came to Morehouse convinced that I was capable of wrecking shop. As a Black man whose father and two older brothers had also gone to HBCUs, I was eager to see how I stacked up against my peers from all over the nation, let alone from countries such as Ghana and Trinidad, where Morehouse also recruits. Instead of my prior school experiences, where I was one of two or three Black students surrounded by utter classroom caucacity, I was amped to step into the biology department's main lecture hall in Nabrit Mapp McBay and see nothing but close to a hundred faces of Black men, all there for the same expressed academic purpose. I sat in the front row because, well, if it was a Goodie Mob concert, I would have tried to get as close to the stage as possible, and wouldn't performing well in a core science class impart substantially more value to my premedical track than Bankhead Bouncing ten feet away from CeeLo and Big Gipp?

The other brother just a few seats away in the front row was Christopher. Not only did he also sit in the front row, he was annoyingly the first one to raise his hand whenever Dr. Clark asked a question. His hand spent more time over his head than resting against his loose-leaf binder, scribbling down notes. What's more, Christopher's answers were almost always right. It was clear to me that this cat was either a formidable adversary or a stalwart colleague. With a hundred students, we both could get As without threatening each other, and we ended up developing a respectful, amicable relationship during our time at the House.

I viewed him as somewhat of a pretty boy: think Christopher Williams, not Christopher Joseph. Being at an all-guy school, however, along with a decision to remain with my high school sweetheart for freshman year despite her going to Purdue, meant that I didn't spend much time over at Spelman's campus anyway. Christopher had carte blanche from yours truly to spit game to as many of the hunnies as he could handle, but we didn't have to compete in that arena because, well, let's just say I knew my limits. But, despite his overachiever, bespectacled pretty-boy visage, it soon became apparent that, even during those undergrad years, Christopher wasn't afraid to march to the beat of his own drum. I vividly recall his shutterbug phase. I couldn't help but take notice of his observational skills—and it's not like

Morehouse had hordes of photography students spilling onto the quad. I mean, we all know about Spike Lee and Samuel L. Jackson, but Christopher was about the only young dude during our time at Morehouse who always seemed to tote a high-resolution camera with a telephoto lens. I suppose that watchful eye he nurtured back in the late nineties gave birth to his keen insights detailing aspects of our society that are often either unrecognized or willfully ignored.

By senior year, he was always playing chess. Did he ever go to class? He was always out on the yard or over at Clark Atlanta's campus scheming ways to sacrifice his pawns for rooks and hoping to checkmate an unsuspecting classmate heartlessly. In reading *The Illusion of Superiority*, his analytical side repeatedly surfaces, such that the reader is left with the impression that Christopher has played these arguments back and forth in his mind, over and over again, as if to refine his opinions and allow him to present a finished product that accurately reflects the patience he has applied to formulating such stances while being able to anticipate counterarguments the way a chess player deftly strategizes to avoid and prevent counterattacks.

At day's end, *The Illusion of Superiority* is a thought-provoking collection of one man's perspective on far-reaching and universal struggles common to every human. He is a philosopher, a scientific theorist, a father, a husband, a physician, a humanitarian, a son, an orphan, a scholar, a patron of the arts, an artist, a student, a servant, a teacher, a friend, and so much more. Christopher is a Renaissance man, and his versatility allows him to toggle back and forth between his numerous identities in a seamless fashion while drawing upon his life experiences to inflect valuable nuances. While only time will tell whether his recent transition to author will be an interregnum before his next career pursuit or a permanent station, his honest appraisal of diverse topics, often deemed verboten, stretch the reader's sense of what is just and unjust, right and wrong, black and white. I assure you that taking the journey with Christopher Joseph in *The Illusion of Superiority* will lead you to a destination rich with contemplation and challenge you to begin changing what you cannot accept in this world around us.

Preface

Thelonius Monk was a bad motherfucker. An Australian might call him a real *Blokey Bloke*, and Black guys like me might say, "he was *that dude*." His playing style isn't everybody's cup of tea, but the jazz legend made his name playing to his own rhythm both literally and figuratively. And it was that courage to go against convention that not only defined his unique jazz piano style, but the reason his music is still listened to and his status as a musical icon endures to this day.

I don't know what the *Illusion of Superiority* will be commercially, or what kind of a book it is really. If I just had to categorize the book, I would call it an autobiographical philosophy book that outlines my journey to identifying alien life. I hope someday readers might feel a Thelonius Monk type brilliance in the unconventional style with which it was written.

Perhaps the most cherished compliment I received about the book so far was from a publishing's company book evaluator who was confounded as to how to market it. His evaluation said frankly, "The author can certainly write... it's an enjoyable read but I find it lacking a central theme and I became lost at times, though not unpleasantly so."

Is this book autobiographical in some parts? Yes. Is it a discourse on my discovery of how to prove the existence of unseen entities, otherwise known as *spirits?* Yes. Is it a self-help book in parts? Yes. Is it a philosophy book? Yes. Do I talk a great deal of other important aspects of society that could make the book feel a bit disjointed to some? Yes, and purposefully so.

Though the verification of the existence of unseen entities, *spirits*, is likely to be how I'm remembered most by society, I am at heart, though a physician and journalist, far more of a *creative* than anything else. As such, I didn't take a conventional approach to writing it. Readers are always on the look out for something new, something fresh, something different. Well, *The Illusion of Superiority* is definitely that, and hopefully readers will feel as inspired by it as I did writing it.

When I realized the analysis of shamanic water was one of the keys to verifying the existence of unseen entities scientifically, it created a series of dilemmas for me. As a married father of three, and a practicing physician, my obligations as both a father and a physician made it impractical for me to just leave medicine and start a phD program somewhere. The financial sacrifice of doing that with a child nearing her college years, and the years of having to study things completely unrelated to my discovery, made such a course obviously not the best fit for my family's needs and my personal research interests. I don't come from money, and haven't made enough of it in my career to dedicate the financial resources needed to do the type of extensive research required to verify through multiple avenues the existence of these entities. As a result, my research to this point has been comprised of solely rudimentary studies with shamanic water. I did what was in my means, and I wrote articles for my blog on www.TheWorldHidden.com, and two books about the subject in hopes such might provide the financial means, and engender public interest and private investor financial support, to assist in what is obviously one of the most significant scientific discoveries in history.

By word an example, I've tried to show my family, my youngest child especially, that science isn't everything, and so included in this book, what I consider perspectives on society that are as important to me as the verification of alien life. That is the main reason for the unique composition of this book.

Is it a cohesive book with a singular theme flowing through every chapter? I think it is, though perhaps not obviously so. *The Illusion of Superiority* has affected humanity in every field of human endeavor for centuries, including delaying the identification of the existence of alien life, so a discussion of multiple aspects of society is essential to have a greater understanding of the diverse ways this illusion still afflicts human society today.

So I invite you to embrace *getting lost while reading it*, and hopefully not unpleasantly so, as I've tried to say nearly everything I wanted to say about society in this book, hence its length and diverse array of chapters. It is my sincere hope my readers will enjoy this book, for what writer doesn't hope readers enjoy her or his work. Is the book unique? Yes. Is it well written? I'd like to think it is. Will there definitely be something that *flips the reader's wig*? Absolutely. So, journey with me down the rabbit hole of *The Illusion of Superiority*, and share your own thoughts once you reach the other side, as I welcome criticism. And hopefully, just maybe, a little praise might come for what I think is a most uniquely intriguing book.

Introduction
Bob Marley Saved My Life

Bob Marley saved my life. I'd seen the cruelty of life in the grotesque ways many people have, and much of what I saw and lived through scarred me. But it wasn't just my own suffering; the pain of others, seen first through the lenses of my Nikon 35 mm as a photographer, and years later as a physician, forever changed me. But I count myself all the better because of it, for few things ripen a heart faster or more earnestly than hardship. Adopted at a young age by an African American couple with impeccable moral values and character, I was convinced by the unconditional love they and others showed me throughout my life that true love and kindness still exist in the world. Without such, overcoming the disadvantages of the circumstances into which I was born, particularly my early life as an orphan, would never have been possible. In a very real way, and in a manner that perhaps could not have been taught more poignantly, I learned *family are those who love you and whom you love*—not just those who share your blood.

  I've never had a close relationship with anyone who shares my blood, yet I've known a life rich with beautiful friendships and loving relationships. Now, I have a family of my own, with children and a wife I aspire to give at least the same kindness and compassion I received from people with no familial obligation to care for me. Fairness dictates that I at least make every effort to cultivate the love so freely and unconditionally given to me through no merit of my own, and why wouldn't I try to reciprocate the extraordinarily benevolent karma some would say I've received.

When I was twenty, I fell into a deep melancholy—*but Bob Marley saved me*. As I grew to understand the world and its suffering, I grew to hate the greed, apathy, and selfishness that's led to the suffering of large segments of humanity for millennia. The callousness of perpetual cycles of commercial and political power intentionally applied to disenfranchise and abuse others, and the willful persistence of these destructive patterns, despite known environmental and societal harm, demonstrates the depth of the blatant injustice that permeates society. It justifiably stimulates outrage and calls for change throughout humanity. I found refuge in Bob Marley's music as I tried but failed to resolve the harsh realities of humanity's wanton history of abuses upon itself and the natural world. Marley's lyrics, with a bond I'd never genuinely shared with anyone else, convinced me he knew exactly how I felt. *Fighting for a better world is worth it.*

I felt overwhelmed from an unrelenting frustration, a frustration common among humanity, especially for those disturbingly aware of being trapped in a system of exploitation and injustice with little to no power to effect immediate change. I hated the world and the series of inept and unscrupulous rulership that's been the pattern of humanity's existence with rare exception. But I loved humankind, and I'd seen throughout my life the beauty and strength of the genuine love and kindness still present in the world, though not always as easy to find as the suffering so prevalent. Songs like "No Woman,

No Cry," "Natural Mystic," "Crisis," "So Jah Seh," "War," "Waiting in Vain," "Jammin'," and other Marley classics filled my mind and heart, encouraging me and reminding me I wasn't alone. Others also want something better for humanity and have taken up their yoke in that fight.

By the time I was twenty-one, I'd achieved nearly everything I'd worked for my whole life. I'd earned scholarships to both college and medical school, was attractive, dated beautiful women, and was popular and well liked. Still, I felt an emptiness inside I cannot put into words. I grew consumed and angered by the suffering I saw and experienced around me. I thought, "Is this the best our species can do? What is my place in this world of madness? Is there really anything I can do to make a difference?"

I was a senior at Morehouse College in Atlanta at the time, essentially coasting through my last year since I'd accumulated enough credits to nearly graduate by the end of my junior year. *But I was tired.* I was burnt out from the rigors of a heavily concentrated science curriculum and no summers off. Each of the three previous summers I'd spent working, first at the EPA and then at two universities doing basic science research to bolster my medical school application. I needed only about fifteen credits to graduate after my junior year, so I spent the year engrossed in photojournalism and writing.

I'd bought an Ansel Adams book, *The Camera*, when I was eighteen, in the summer after my freshman year, and read it from cover to cover in just a few days. A few weeks after finishing the book, I bought my first 35 mm film camera. I walked into the camera store knowing exactly what I needed—a Canon Rebel or a Nikon N60. The Nikon was heavier, and its gunmetal grip felt more robust and comfortable in my hand, so I went with it. And I was hooked. I started taking pictures everywhere I went. I'd found a way to show people what I saw when I looked out at the world, and I thought perhaps my work might make a difference to someone somewhere.

From the time I was a child, I'd felt impelled to become a physician. I felt a commitment to and passion for medicine and understanding the human body effortlessly. Medicine grabbed a hold on me when I was just four or five years old, as I watched my grandmother injecting herself daily with insulin. Even at that early age, I was naively confident I would someday, somehow, be able to cure her diabetes as a physician. Years later, the more I wrote and spent time photographing my surrounding community in Atlanta, I realized medicine didn't hold all the answers to the world's complicated and diversified problems. The suffering I saw was not the result of a deficiency of knowledge, technology, or health care—but of morality. I saw and felt the power of photojournalism and photography in the works of iconic photographers like Roy DeCarava, W. Eugene Smith, Henri Cartier-Bresson, Gordan Parks, Chester Higgins Jr., Phil Borges, and others. Their work enraptured me, and I spent hundreds of hours in the AUC's Woodruff Library studying their photographs, hoping to capture the essence of what made their work inimitable.

I became addicted to photojournalism and photography and carried my camera everywhere I went. I woke up early every day, even on the weekends, and just drove. I'd go to downtown Atlanta, Piedmont Park, Auburn Avenue, the West End, and other areas around the city, especially near the college library, and I'd shoot whatever spoke to me. One day I saw a couple of kids playing near a sculpture on the north side of the library lawn. Two handsome but obviously impoverished young boys, no more than ten years old, were climbing through the statue resting there on the lawn. It would have been a good shot, but I didn't photograph them for many reasons, not the least of which was an awareness of the need to have an impeccable reputation to do the kind of candid and intimate photojournalism I wanted to do. I'd learned through experience the importance of humility and compassion when approaching strangers for photographs, and nobody welcomes some random guy aiming his camera at kids he doesn't know, especially in the ghetto.

Accounts like Gordan Parks' iconic photographic experience with Ingrid Bergman also taught me that, when ethics and empathy bid you put down your camera, any opportunity lost through thoughtful consideration of another will be replaced in time through patience and goodwill. Bergman, an A-lister in the midst of scandalous marital affair by 1940s standards, was filming a movie on Stromboli Island with her then lover, director Roberto Rossellini, for whom she'd left her husband and child. Parks had been sent specifically by *LIFE* magazine to obtain a photograph of the controversial couple embracing—but he was not the only photographer given that assignment:

In *Voices in the Mirror*, a memoir, Parks wrote that he was well aware of the bind that she created for him. LIFE's editors wanted what the world wanted — "that unguarded moment of passion." "If the moment arrived," he wrote, "either LIFE or the melancholy lovers would suffer."

When the chance to betray Bergman presented itself, Parks let it pass. After a day's filming had ended, he entered the darkened set and found her and Rossellini holding each other in an intimate embrace, one that was "comforting rather than... lustful." It was the kind of "unguarded moment" that he was on the island to capture. He began to raise his camera but immediately put it down and left the room, hoping that neither of the lovers had noticed his presence. "The moment that slipped away," he wrote, "was undeserving of betrayal."

Bergman had noticed, however, and she was grateful. A few days later, she invited him to follow along as she and Rossellini walked along the island's shore. On their own terms, they gave Parks the photographs that his editors wanted.

— John Edwin Mason, "Gordan Parks"
Photographer

The boys headed off west, up the block adjacent to the library, shortly after I arrived, and I headed off in that direction, too, since I didn't know the neighborhood well and felt like exploring. I walked two blocks up, but the boys had walked off out of sight as I approached a group of men playing chess on a porch. I put my camera down immediately. The thrill of a game was more interesting than any photo at that moment, as I hadn't played a good chess game in months. The itch for a good game of speed chess, two to five minutes usually, never fades too far from me. It's an itch I've had since I was a teenager, when my brother first taught me the game.

The men were sitting in front of a single-story house with a large porch. There was a small wooden chessboard with worn plastic pieces to the porch's front right side. Two poorly upholstered wooden chairs held two chess players, both Black men in their fifties. Behind the table, flush with the house, was an old white fabric couch, where three or four other men sat drinking beer. I spent the next several hours sitting with them, talking shit and taking turns playing chess. I was rusty from not having had a quality opponent for a while and lost more than I should, but I still managed to win a few games. As we played and talked, I got to know the men, and, over the next few months, we all became close.

The boarding house manager, a man who would become both a friend and mentor to me, sold weed to several people who came and went as we played. In time, the father of the two boys approached him while we were playing. I knew he was their father immediately, as the boys were the spitting image of him. He was mid-thirties, a slim but fit light-skinned Black man with long dreads. I asked him, "Would you mind if I took a few photos of the kids playing?"

"Sure," he said—and that began a relationship with the men of that boarding house, the boys' family, which included six other kids from ages two to sixteen, and several other local families.

The father of the boys was a struggling crack addict. He had a small two-bedroom home, catty-corner from the boarding house where the men played chess, and all ten members of his family lived in that home. For the next several months, I spent time photographing them and the community around the college and throughout Atlanta. It was one of the most productive times of my many years as a photographer. It was also the most significant period of growth in my life as a person.

The time I spent photographing the impoverished of Atlanta changed my perception of poverty and the poor forever. I hated the inhumanity of poverty and the circumstances that often lead to children suffering through no fault of their own. I hoped that my work would bring light to the efforts still needed in impoverished communities throughout the US.

But I also saw the beauty of the poor. It was a pattern I would repeatedly experience traveling and photographing the world around me over the years that's followed since my early time as a photographer and journalist. The honesty, sincerity, and kindness I experienced moved me in a way I've never forgotten and challenged my conception of what comprises true happiness and satisfaction in life. I also became a more compassionate and understanding person and chose my friends with scrutiny for their character much more intensely after that experience. I learned to cherish genuineness in my friends above all, and I saw the world, in all its shades of gray, in ways I never did before. I'd come face-to-face with prostitutes, gang members, drug dealers, drug addicts, honest working-class people, and others with lifestyles I'd never known firsthand, in the most intimate and personal of settings. Those experiences also showed me that loyalty, honor, dignity, and morality are often present in enormous measure among those most often assumed to lack such.

How did an uneducated ex-convict who sold weed become a friend and mentor to me? What he was above all was my friend. When I sat down at the chessboard, I didn't know or care who he was, what he'd done, or how he lived. I was there to play chess. I'd had trouble finding a good game on campus, and, when I saw the board, I just wanted to play. Any real chess player who's spent any time playing speed chess knows some of the greatest speed players are people who, on the surface, might not seem intelligent or virtuous. But I've played on the streets of New York, Atlanta, Boston, Philadelphia, Cuba, St. Kitts, and Camden, and in many other impoverished communities. Some of the strongest players I've ever faced had missing teeth, were criminals, smelled of alcohol and cannabis, were homeless, and were otherwise unsavory on the surface.

As we played, and my visits gradually became more common, we talked extensively and got to know each other. I saw a side of life I'd never known intimately before being raised by my sincerely religious and virtuous parents, and the landlord was full of the wisdom a life of regrets and mistakes often instills a person with. He went to prison as a young man in his twenties, a heroin addict trapped in the cycle of violence and crime, and jail helped him get sober. He never used heroin again. He spent his time reading in prison and was a well-read, brilliant man in his fifties when we met. He had a strong disposition and commanded the boarding home tenants' respect with shrewdness and the presence only a man experienced with dealing with criminals possesses. We spoke of his time in jail, his regrets about his relationship with his estranged daughter, and we shared a fondness for religion, sociology, philosophy, and, of course, chess. He saw some greatness in me that even I didn't see at the time. He loved my photography, and he respected my intelligence and the genuine affection I felt for, and how I dealt with, people of considerably lower social status and educational level. He was convinced I would become a man of great consequence in society in time and encouraged me to embrace the greatness he was convinced I was destined for. He died about a year after I started medical school, but his friendship and our conversations on that small porch remain among my most cherished memories.

Still, it was Bob Marley who remained my closest constant companion. As I drove to and fro on my photographic excursions, his music and my camera became the inspiration for a new journey in my life. Songs like "Waiting in Vain," "No Woman, No Cry," "Is This Love," "Pimper's Paradise," "So Much Trouble in the World," "War," "Trench Town Rock," "Natural Mystic," "So Jah Seh," and other Marley classics became my greatest comfort and the rocket fuel for the most creative time of my early photographic and writing life. It was a journey I'd never imagined making before I was senior, one that manifested organically from the inner turmoil of trying to make some sense of the broken world around me. My senior year at Morehouse became a beautiful time of personal growth and artistic discovery..

As graduation approached, I had an onerous decision to make: continue on my new journey as a photographer or accept the full scholarship to medical school I'd worked my entire life to earn. A surprisingly successful exhibit of my work in a local Atlanta gallery made the choice an even more difficult one. I talked to my parents, who were still convinced medicine was the right choice for me, but conflicting thoughts about my future lingered. The decision to accept the scholarship and head off to the Northeast to start medical school I made reluctantly. Deep inside, medicine was no longer the love of my life, but I reasoned I could photograph anywhere. The potential instability of a career as a photographer, the fear of disappointing parents who'd made numerous sacrifices for me my whole life, and, frankly, my own fear of failure made chasing my artistic dreams untenable for me at the time.

I survived the first few years of medical school as an above-average student but just a shadow of the type-A, straight-A perfectionist who characterized my high school and college academic performances. My heart wasn't in it entirely, but I still did better than just get by.

I wasn't the type who could just chill and study at the very end before the test. I had a classmate with a photographic memory who did that regularly. That wasn't me. I needed to study every day and get the most out of those sessions. On the weekends, when I could escape the drudgery of hundreds of pages of weekly reading for a few hours, I'd walk or drive around with my camera in Camden, New Jersey, and Philadelphia for a few hours, hoping to recapture some of the magic in my art I'd first found in Atlanta. Sometimes I got lucky—most times, I didn't. But at least the camera remained by my side.

Such became the pattern that would define my life and stick with me for the rest of my time in medicine. I'd give 85 percent to medicine and the last 15 percent would be for me, for writing, for studying religion and philosophy, for photographing, and for playing music. That 15 percent has always been the medicine keeping joy, freshness, and change in a life full of excessive hours spent serving others. On my vacations, I'd travel somewhere new, make friends, take pictures, and reconnect with the natural world through hikes and some time usually spent on the water. Bob Marley was still never too far away, and his music, and that of others, stayed my constant companion on my ongoing adventures, sometimes on foot, but mostly by car, in new places like Montserrat; St. Kitts; Miami; Philadelphia; Baltimore; Seattle; New York; Puerto Rico; LA; the Jersey Shore; Ocean City, Maryland; and countless other big and small towns on road trips. Everywhere was a new landscape with new peoples and experiences to savor and capture, and new cultural experiences to discover.

Eleven years after I moved to New Jersey, I decided to move again—but this time to the Southwest. When I moved to New Mexico, another beautiful transition happened in my life, one that I hadn't anticipated or even considered when I first made the decision to move. For the first time in my life, I got a chance to spend significant time around the Indigenous communities I'd previously only read about. When I married my wife, I also grew to understand Cuban culture and the significant population of traditional healers there. It was a combination of serendipitous events that would have considerable import on my understanding of shamanic cultures around the world, as the similarities between Native American shamans and Cuban *santeros* and *curanderos* became obvious to me. Those commonalities, along with the understanding of the occult and spirits that my study of world religions instilled, led me down the path that resulted in *The World Hidden*, my first book. I began rethinking the way I viewed alien life within the context of the reality of the existence of unseen entities, which became unequivocal after my initial interaction with a shaman.

When I met with a genuine shaman for the first time, she told me things that my closest friends, and even my wife, didn't know about me. She also told me things I didn't know about my family at the time but have since confirmed as being accurate. I realized immediately that her ability was not something customary for humans, and she confirmed that her clairvoyant insight was due to an ability to interact with unseen entities, commonly referred to as spirits by many cultures. This experience lit a flame inside me to understand shamanic cultures globally and to relate that knowledge to others, for I realized immediately that the existence of unseen entities represented a significance of inestimable value for society and that alien life was not something humanity need search the universe to find. It has, in fact, been surrounding humanity for millennia, though in a form and manner the overwhelming majority of humanity in modern times poorly understands and fails to recognize.

It's been twenty-five years since I held my Nikon N60 for the first time, and writing about and photographing the world has remained the seasoning that keeps my life balanced from the rigors and challenges of medicine and family. Though not always the case in the past, I no longer regret my choice to become a doctor. Medicine has provided a life of stability for my family and me; despite the challenges, I've managed to remain an active writer and photographer.

My decision to become a doctor, along with my experiences visiting and documenting Indigenous people and impoverished communities, left me considering many things about life and happiness in general. *The Illusion of Superiority* was born in recent years as I reflected on my choices, successes, and failures, and my understanding of Indigenous philosophies and views on life. Indeed, my life in medicine has been stable financially, but was I a happier person than I would have been if I had chosen photojournalism and writing full-time? And what does it mean to be satisfied and content?

Such questions get to the heart of me finding meaning in my life, an endeavor that has grown in significance for me as I've aged and now consider the lives of my children and their futures more preeminently. The ephemeral essence of life is such that the consequences of some paths and decisions have lifelong ramifications. For example, the decision to pursue medicine took me away from my family for years, with over a decade of my life spent in universities, studying the best years of my life away. Reflections on my own path and choices made me rethink my opinion about the value of material wealth, a career, family, love, and many other things. Those reflections, combined with observations over the years from practicing medicine, studying, visiting, and photographing various Indigenous peoples and other impoverished and affluent communities, refined and challenged the prejudices and preconceptions Western civilization implanted and indoctrinated in me over decades of study and matriculation through American society.

My wife grew up in a large family in Cuba. My father the same, though in Alabama as the youngest of ten children. Many of the communities I've grown fond of feature generations of families living in close proximity. I've known and photographed migrant farmers, including some illegal immigrants, in New Jersey, who spent six months of the year in the United States working and living in commune-like dormitories, sharing food and resources while working and then going home to live six months in countries such as Mexico, Guatemala, and Honduras, despite the difficult journey. I've had other friends and acquaintances, some extraordinarily affluent, who've suffered both similar joys and unique difficulties in the same manner as their far more materially bereft counterparts. Who really is better off? Who is happier? Who is more satisfied with life? Is their one path truly best in life to follow?

My study of Buddhism, Islam, Christianity, Indigenous teachings and philosophies, and personal experiences all convinced me that cultivating relationships is the most important pursuit in life. For a devout religious person, for example, the pursuit of a relationship with their creator would rightly be considered the most important pursuit in life. For Muslims and Christians, for example, their prospects of everlasting life rest solely on the quality of their relationships with their creator. Many religious texts implore such a course. For one not inclined toward belief in God or religion, maintaining healthy relationships often rests as the foundation for attaining and growing material prosperity and for satisfaction and contentment in life, as no degree of material prosperity brings genuinely loving and healthy relationships with family, friends, and romantic interests. And no person's life, regardless of whatever external achievements or wealth a person acquires, can be a satisfied and contented one without loving relationships in it.

    Many traditional Western philosophical thoughts on purposeful living, success, and achievement focus exclusively on the individual alone from the natural world. Stoicism, epicureanism, existentialism, relativism, all prominent philosophical traditions, give attention to the individual as an entity set apart from the surrounding natural world, and prominence is not given to the import of ecosystem health as a value to be instilled and encouraged. Nearly without exception, Western civilization has focused on superficial measures of success, and the subsequent estimation of a person's perceived value is rooted in measures that in themselves speak nothing of society's, or the natural world's, well-being as a whole.

As a species, we are united by the universality of our shared dependence upon Earth for our survival, a bond we also share with every other nonhuman living thing on this planet. Yet the dominant philosophical and educational priorities of Western society have never shared harmonious coexistence and well-being with the natural world as the apex of human achievement. Western philosophical traditions, by their focus on individual achievement, have produced a global system that is exactly in harmony with what should be expected from individualistic focus, that is, the human species plagued by environmental decay, broken families, rampant inequality, and the proliferation of every manner of injustice and abuse toward oneself and others, as individuals place the achievement of individual material prosperity above every other value.

The philosophical traditions of numerous Indigenous peoples stand in direct contrast to the individual focus of many Western philosophical traditions and practices. Nearly without exception, Indigenous communities place family and community well-being above all, and there is no concept of success or fulfillment outside of a commitment to harmonious coexistence with the natural world and the community with which one shares codependence within the delicate web of life. There are a great number of practical and insightful principles found among various Indigenous philosophical traditions, and their application is vitally needed in the modern world. As a species, Indigenous principles hold the key to our indefinite survival, and a consideration and implementation of their profound merit lies at the heart of any hope modern society has for a more egalitarian and benevolently prosperous global civilization.

The adoption of patterns of behavior that threaten humanity's progress as a species warrants inclusion into similarly solemn consideration, for, in the words of James Baldwin, "not everything that is faced can be changed, but nothing can be changed until it is faced." There are aspects of modern civilization that make genuine liberty and the pursuit of happiness difficult for large segments of humanity, within and outside of industrialized and non-industrialized nations. Discussion regarding aspects of governmental failures in the exercise of its authority is not only a righteous endeavor but a necessary one, as the practice of government in all its forms has been as much a driving force for societal change, often from revulsion of tyranny by its citizenry, in as much as it has contributed to its citizenry's protection.

So, we begin this journey, first with consideration of the individual within the context of their place in society and, later, a discussion of the macrocosms of society itself. At every level, humanity's self-governance has, in practice, revealed enormous room for improvement. It is with a spirit of hope for the reconciliation of humanity with itself and the natural world that I pen these words, for there is no wisdom in finding illness without hope for remedy, nor nobility in finding disease without offering insight for a cure.

Chapter One

The Discovery of Alien Life
Reflections on Unseen Entities and *The World Hidden*

The intuitive mind is a sacred gift and the rational mind is a faithful servant. We have created a society that honors the servant and has forgotten the gift.
— Albert Einstein

Intuition is a very powerful thing, more powerful than intellect.
— Steve Jobs

What the hell happened to CNN? The network used to be way more toward the center when Ted Turner was the owner. Fox has always been hard to the right, that's to be expected from them, but CNN has become the anti-Fox, a twenty-four-hour anti-Trump network now it seems, with little more mentioned than anti-Republican politics and the coronavirus. Hello McFly?! Hello?! Other important things are happening around the globe every day, and yes, they deserve some quality news coverage too. Don't get me wrong, I still enjoy the network, I do. Fareed Zakaria is a very fair journalist and has one of the best shows on television. The CNN anchors are still some of the best on TV to me, and Chris Cillizza is one the best writing journalists I've ever read. I usually watch an amalgam of CNN, Fox, and Al Jazeera a little every day, hoping to piece together the puzzle of what the truth of what's happening in the world really is.

Journalism is a beautiful and incredibly powerful medium. In recent years, particularly in the age of Trumpism, the media has come under significant attack from both the political Left and Right—and not without merit. It's difficult in this era of corporate addiction to TV ratings and extreme political polarization to find pure and unbiased journalism. Information gathered honestly and with integrity and stories focused on global societal impact, told artfully without some hidden agenda, are indeed rare gems. Such is the modern journalistic environment, and it can be easy to dismiss this beautiful medium, an under-appreciated art form in my opinion. Still, there is incredible elegance and power in telling the truth in a way that moves people with the essence of honesty. And if you don't believe journalism can be art, take a look at a few episodes of Anthony Bourdain's *Parts Unknown* and we'll see if you still feel the same way.

How do you convince people of a truth they cannot see? That was the central question I asked myself after I became certain unseen entities are real. The world witnessed the power of journalism as a medium when we watched the death of George Floyd play out on televisions, phones, tablets, and computer screens around the world. The documentary footage of Floyd's cruel and callous murder sparked protests throughout the US and other countries. That is the true power of journalism.

The video of Floyd's murder couldn't be spun to paint a different picture. There was no hidden guise within the video and no way to convince the world that what it witnessed was anything other than what it was: a heartless murder orchestrated by a member of the police.

That is the power of journalism at its heart. Documentary video, as photojournalism did before the advent of video, cell phones, and social media, elevated journalism to a uniquely more powerful plateau. If the famous often-quoted mantra "a picture is worth a thousand words" is true, then how many words is documentary footage worth?

When I wrote *The World Hidden*, I knew the book and its accounts would come under intense scrutiny. I was sharing accounts that appropriately should be met with great skepticism, as many of the accounts stretch credulity. I was postulating the existence of intelligent alien life-forms that cannot be seen with the exception of a few among humanity, something that even I would have difficulty believing had I not personally experienced the power of genuine shamans. That is a declaration of unquantifiable significance and consequence—not just for science but for religion and humanity itself. Such a proclamation should be met with the highest degree of scientific scrutiny, and every effort should be made to verify the accuracy and veracity of such accounts.

The World Hidden was born principally because I lacked the means to do what I really wanted to do when I first understood the significance of unseen entities and their activity with humankind. Initially, I wanted to tell the story of Indigenous healers through a blend of documentary filmmaking and scientific study and by monitoring the biometrics of these healers' brains and bodies. Brain analysis while in transition (a term used to describe when an unseen entity is directly interacting with someone), radiographic imaging (PET, MRI, and CT scans), the analysis of patients' blood chemistry before and after treatment, DNA analysis of generations of healers from various locales globally, and complex and sophisticated analysis of the water from these healers' altars form the principal foundation of the research work I want to do. After forming a friendship with a shaman and visiting with her over an extended period, I noticed that the water in the wine glass she used to perform consultations changed. It took on a unique opaqueness, an apparent alteration in its chemical composition that does not happen when water simply sits in a wine glass. The morphology of the water in the glass also appeared unique, with a perfect sphere just below the surface of the water developing over time.

Water is a commonly used tool among many shamans as a means for interacting with unseen entities during periods of clairvoyance. Water is often thought to sequester and hold these entities. This is a nearly uniform custom for many shamans, particularly so for *santeras* and *curanderas*. In time, I grew convinced that a simple but thorough analysis of this water would conclusively, scientifically reveal the truth of these unseen entities; it is also relatively easy research to do if one has access to cooperative genuine shamans and the financial means to do so. So, I began my work in this area and in a short time saw the accuracy of my initial assumption. Shamanic water is the principal way to demonstrate the reality of these entities' existence.

Researching Unseen Entities

Unseen entities interact with the physical world principally in three ways, and it is through examining these three ways the evidence of their existence can be established beyond all doubt.

I. Analysis of shamanic water

Unseen entities are sequestered in the sacred space of shamans and others with the ability to interact with them through the use of water. This water is then used to perform consultations with clients seeking assistance. Nothing can effect a change in the physical realm without leaving some trace of its existence. Unseen entities are no different in this regard.

The water of shamanic altars is altered in verifiable ways, principally because these beings are comprised of a type of energy that modern scientific study has, as of yet, been unable to fully identify, assess, and understand. Because of limited financial resources, I've been unable to complete all the various types of analysis of this water I'd like to pursue, but the morphologic change of this water is the most obvious change that can be readily observed. Study of the *why* and *how* of this change is complicated and demands a multidisciplinary approach. Variations in the conduction of electricity, variations in light and laser transmittance, the effects of temperature fluctuations on these and other factors, including spectral imaging, electron microscopic imaging, and other expensive and complicated physical assessment tools useful in scientific evaluation, such as gas chromatography and mass spectroscopy of the water before and after shamanic intervention, form the foundation of the multitude of research experiments that must be completed on shamanic water to give a more complete picture of how this water is altered by the presence of these entities.

II. Biometric analysis of shamans in transition

The physical and chemical examination of water alone is not enough to establish unseen entities are responsible for the change in shamanic water. It is vital that three levels of experimentation occur simultaneously to establish beyond all doubt that the change in water, the change in shamans, and the change in individuals in consultation with the shaman are all related and interconnected.

Transition, a term often employed to describe the period when shamans channel unseen entities into their bodies, is characterized by changes to both an individual's motor and sensory perception and control that can be monitored, measured, and evaluated. When shamans enter transition, they lose conscious awareness and cannot remember anything that these entities say and do during this period. The period before conscious awareness is lost but when the process has begun of the entity gaining control of the shaman's faculties is characterized by physical changes in the shaman. Frequently changes occur in the motor control of the tongue and the speech of the shaman is affected. Such changes in mentation and physical muscular changes likely will have corresponding EEG (electroencephalography) changes. There are also likely neurohormonal changes and other biometric changes that can be observed with spectral imaging, likely linked to the changes observable in the water used in shamanic rituals.

Another interesting pattern described by the large collection of shamans I've interviewed is what they describe as a period of awakening after the entity is allowed to interact fully with them and gain control over their faculties. The immediate period after transition is characterized by a change in the shaman's visual perception, where colors and other physical sensations are enhanced. Colors appear more vivid, for example. The long-term effect of transition is that shamans are from the period after initial transition able to see the entities the same way they are able to see other physical objects. This change suggests that transition alters the brains of shamans in some permanent way that perhaps alters the visual spectrum their brain is able to perceive. This is an alteration that must be assessed and studied to be fully understood and comprehended.

III. Biometric analysis of patients

The third and final crucial component of analysis that must be incorporated and coordinated with the simultaneous analysis of both shamanic water and shamans is the biometric analysis of patients interacting with shamans. It has become obvious to me that unseen entities are comprised of, and operate principally through, the use of a type of energy that as of yet humankind has little understanding of. That energy is transferred through shamans during the work they accomplish, and it will have some measurable consequence in the physical world that I believe is observable in shamanic water, shamans, and those toward whom such energy is directed.

Many types of healing feats are reported by the shamans and others I interviewed in *The World Hidden*. Biometric analysis, spectral imaging, and radiographic study prior to, during, and after shamanic treatment intervention comprise the principal ways to monitor patients treated for conditions with objective measures that can be readily monitored. Autoimmune markers; tumors, particularly inoperable ones; hypertension; and other conditions offering concrete observable metrics will show whether any therapeutic intervention has any consequence on such conditions. Other psychiatric conditions such as PTSD, drug and alcohol addiction, schizophrenia, and depression are all predominantly subjectively assessed, in the sense that what patients describe they experience largely provides guidance about the quality and effectiveness of any given treatment, and these should also be included in evaluation. Various assessment tools are commonly used in research studies to measure the effectiveness of various treatment modalities. Psychological conditions should also be included in research studies, for perhaps in no greater field of medical endeavor is the need for innovative and effective mental health treatments as great as ever. I suspect shamans and unseen entities work primarily through the manipulation of the energy of intention, the same energy that allows us to sense when others are staring at us though our back is turned, for example.

There were many things I saw with my own eyes and others that were relayed to me that I intentionally left out of *The World Hidden. Words are just not enough sometimes.* Many of the accounts I shared I knew would be hard to believe, and I didn't want to challenge the credulity of readers beyond a certain point. Many of the book's contributors, including my wife, were exceptionally candid and forthright in sharing their experiences with me for *The World Hidden*. I knew many of the experiences I excluded from the book would simply be met with ridicule, even beyond how the book is already scrutinized. My wife recalls, when she was a young girl, watching her grandmother make it rain and stop thunderstorms by just putting an iron hammer on the ground. I've personally observed the manifestation of shamanic shapeshifting, and there is simply no number of words, no matter how artfully crafted, that will convince people of the reality of such things without showing them. Documentary filmmaking combined with organized scientific research is the only reasonable path forward to demonstrate what shamans and others are able to do through their interaction with unseen entities.

    A few years before I had my transformative experience with a shaman and became convinced about the reality of unseen entities, and not long after I moved to New Mexico, I pursued a couple of unique entrepreneurial endeavors. I'd written the rough outline for a sci-fi story that featured an Indigenous couple as heroines. I loved the story outline but let it linger for a while before writing it. At the time, I was working in an office practice and in urgent care, and I would play more video games than I should on the weekends and at night after my long shifts, during which I sometimes saw sixty patients or more. I'd recently met a patient who was a genius computer programmer, and he'd developed some of the most famous video games of all time and other simulation programs for the military. We talked and got to know each other. He loved the idea of the story, and I presented him with some novel ideas for gameplay and replayability that would enhance the game, and we got to work. I further developed the story and had storyboards made. We recruited local artists to help make 3D models of characters and environments. The artists were volunteers, students from the local university and community college, and though they were talented, it's often difficult to get the required quality and rate of production from students without professional experience. Eventually, the project stalled.

  A few months later, I began thinking about something else the programmer and I could work on together; this time, I chose something educational. I had an idea for interactive textbooks and medical training software that would blend augmented and virtual reality to make an innovative teaching tool for medical students. The idea would be to train students on how to think like a doctor and to present training scenarios for complicated medical procedures that are otherwise only learned on actual patients. The programs would expose medical students to common emergencies, for example, how to deliver a baby, put in a central line, decompress a pneumothorax, intubate, do nerve blocks, and other common procedures and scenarios most doctors should be comfortable handling. Heart attacks, aneurysms, strokes, GSWs (gunshot wounds), vehicular trauma, and other frequent emergency and intraoperative complications would be part of the training. The students would handle these scenarios repeatedly in virtual reality and train for procedures with a combination of virtual and augmented reality with cadavers and dummies so that when the time came to complete the procedures on actual patients, they would have considerably more familiarity with the procedures and scenarios customarily students and residents get little to no training for prior to real-life situations.

The point of the innovative program was to reduce the likelihood of error, which, in my opinion, is a significant problem with medical training in that much of what must be learned is typically, predominantly done on actual patients. This presents obvious problems, and anything that can reduce patient risk and enhance student learning is worth at least considering. I presented my idea to a local medical school, and it was universally accepted as practical and useful, but because of the significant cost associated with developing the program, it was dead on arrival. I'm not the sort who has a cadre of wealthy venture capitalist friends on hand to front me the capital for a product that would require millions of dollars of upfront investment to develop, so though the idea was good and practical, it also stalled. I placed it on my list of failed inventions and other ideas that went nowhere because of a lack of capital.

Looking back, the disappointment I felt from the experience of the failed video game and medical training program helped me realize the importance of not aiming too high initially when trying to achieve something of great significance. I refused to make the same mistake I made approaching unseen entities as I made with these and other projects. I deftly know my way around a camera, but I'm no cinematographer. The documentary I want to do is international in scope and sophisticated in nature, with medical imaging, biometric analysis, and basic science research all encompassed. I completely outlined the process and thought deeply about how to proceed; this time, however, instead of shooting for the moon, I chose to start with just writing a book, a far more realistic goal. I was driven and had the help of many benevolent shamans and others they'd helped. I also realized that I needed to let the people speak, which I believe is where the true power of *The World Hidden* lies.

The World Hidden's appeal, in my opinion, is that it is principally a work of journalism. There's some modest commentary by me, outlining the societal ramifications of the work shamans and others who interact with unseen entities do, but the book's true beauty is in the accounts of the people who shared their stories with me. To my wife's dismay, I simply let the people speak and didn't try to craft conversation and narratives too elegantly. There's tremendous power in preserving the authenticity of a person's experience. My time as a photographer and photojournalist taught me that, and a lot about dealing with people in general. A camera is an intrusive instrument, and honest documentary photography requires accessing people's vulnerability.

When I was at Morehouse, not long after my first exhibit debuted, a respected friend asked me if I thought my work was exploiting the impoverished people I was photographing. I looked at him, thought about it a moment, and answered truthfully that it in some sense, yes, it does. I then sincerely responded that the intention behind the work was noble. I was convinced of the importance of showing a side of life that is often left unseen or shown in a way that lacks nuance, showing only the ugly side of poverty. I also gave the people who I photographed a significant portion of the money I earned from the exhibit. There's never been a harsher critic of my work, both ethically and artistically, than I, and my conscience was clean knowing that my intention and my manner of practice showed both respect and gratitude for the individuals who let me capture a part of their lives.

    I've always, and without exception, got permission before I photographed anyone, and still do to this day. I followed the same principles in collecting the accounts shared in *The World Hidden*. I've never shared anything that any person granting me access to their life didn't approve of me sharing, and I never will. In collecting the accounts for *The World Hidden*, the shamans and others who interact with unseen entities, I'm sure, knew that I would respect them in this way and respect any boundaries they put on what I could and could not share. This in no small part contributed to their candor. As a journalist, particularly a photojournalist, the aura you manifest and the spirit with which you engage in this type of intimate work can be felt by those whom you photograph and other observers, even those without the unique hyperdeveloped intuition and clairvoyance of shamans. That aura, one that should display a commitment to fairness and integrity, humility and confidence, lies at the heart of encouraging individuals to be open and honest when sharing things on the record.

For centuries, documented accounts of Indigenous people presented various abilities shamans manifest as a result of their ability to interact with unseen entities. In writing *The World Hidden*, I didn't discover alien life or present anything the world hasn't heard before. For me, the beauty of the book rests in the words of the people who shared their accounts and, in my relating through diligent verification, the veracity of those accounts. Masterpieces like *El Monte*, Eliade's *Shamanism*, *The Book of St Cyprian*, *The Keys of Solomon*, and other ethnographic and occult masterpieces have outlined in far more detailed and elegant ways the depth of ability shamans and others who interact with unseen entities manifest. The variety and diversity of thought regarding the ability of unseen entities is presented far more elegantly in these works, and those who through various means interact with and use unseen entities to perform a variety of difficult-to-believe feats are explained in significantly greater detail.

When I first shared some of my writing with my wife, I asked her what she thought of it. Her cryptic response was, "You write like a man"—not exactly the feedback I was hoping for. I wasn't exactly sure what she meant at first, but the tone of her voice and the look on her face weren't exactly encouraging. I compelled her to explain what she meant. "You're logical and you express yourself clearly, but there's no color, no flamboyance in the way you wrote it. It feels like a journalist telling the news." I couldn't deny it. She was right, and I laughed as I objectively considered her words, but that was exactly my intention. I was trained as a journalist and never really broke that training. My first editor shared my wife's opinion of my writing and had no qualms in repeatedly telling me my writing was shit. She never explained why or exactly what she didn't like about it: She just wasn't a fan. Some writers are master story tellers, others, exceptional wordsmiths, and the truly rare ones like my wife and Jose Martí are a combination eloquence and profundity, whose range is broad and includes everything from political essays to poetry for children. But *The World Hidden* isn't about me, or my writing, it's about the people who honestly and candidly shared their accounts with me, and I stuck with a simple and straightforward tone.

    My wife's is an incredible writer, with a foul mouth and a spectacular sense of humor. She just has a way with words and with people. Her style and way of connecting with people simply cannot be taught. She naturally has something I'll never achieve no matter how long I work on my craft. But she writes to entertain, and I was writing eyewitness accounts trying to match the tone of how the individuals told them to me. We're opposites, my wife and I, both as writers and individuals. When she read my writing in *The World Hidden*, it wasn't her choice of style; still, she was convinced that my work would sell and that I'd become a successful writer in time because of the depth of my ideas and the significance of my observations. In a way, her reaction often illustrates the ability to distinguish something that's entertaining from something that is practical and useful.

The World Hidden is not a book many people would usually even consider. The man who wrote the foreword for the book is a close family friend. As he is a physician and an intellectual with a lifetime of distinguished accomplishments and medical credentials, I asked him to write the foreword because I knew he would not only do a good job but also that he would be completely honest. As a conservative Christian and an exemplary physician, he represents both segments of the populace I felt would be most resistant to the ideas of the book. If he found the concepts and conclusions insignificant, it would be worth hearing his objections, and if he saw the merits of the accounts and theories, I had hope others whose inclination would be to dismiss my findings would also find value in the book.

His response, along with those of other friends and family members, convinced me I had indeed found the middle ground: a book that had significance and one that could be understood and considered by people from all walks of life. I revisit some of those same concepts here, with hope that the obvious merit of my work and its profound ramifications for humanity will touch the minds and hearts of the reader.

Are Unseen Entities Actually Aliens?

Are unseen entities, commonly referred to as spirits, actually aliens? Why have I chosen to refer to them as alien life-forms? For many of the healers I've spoken with, unseen entities are considered far more than aliens, whereas for some they are deities and the rituals done to gain various entities' favor are acts of worship. For others with the ability to interact with unseen entities, these entities are not deities but the remnants of dead loved ones, something left over after we die. Still, for others, they are simply nonhuman intelligent beings, some of which are benevolent and others malevolent. For some, the rituals done to interact and gain favor with unseen entities are simply an exchange in which the entity is granted something desired in exchange for some favor or wisdom the person interacting with the entity is seeking to gain. Rituals to gain protection, healing, clairvoyance, and guidance for a particular problem are routinely done by shamans and others who interact with unseen entities.

I have chosen to describe unseen entities as forms of alien life for several reasons.

1. There is great divergence of opinion among shamans and others who interact with unseen entities as to what these beings actually are. A nonspecific term is helpful when attempting to classify these entities since it is still unclear to me what they are scientifically.

2. Unseen entities show obvious signs of superior intellect and powerful ability and are nonhuman. Those with the ability to see and speak with them report having conversations with these beings that are often diverse and varied.

3. For too long, the existence of unseen entities has been entirely shrouded in the realm of religious mysticism. Religion, television, and cinema have

nearly entirely defined how society views unseen entities, often showing only the destructive and malevolent aspects of some entities, completely ignoring the accounts of shamans and others who interact with unseen entities about their ability to cooperate with these entities for the benefit of others.

4. Referring to what many cultures commonly name spirits as unseen alien life-forms helps bring an objectivity that has been sorely missing from the consideration of unseen entities by society, an objectivity that is essential for sincere scientific evaluation, study, and classification and comprehension of the risks, dangers, and potential benefits that these beings pose.

Because there's never been mainstream acceptance of the reality of unseen entities, there's never been the type of organized and detailed study and research these beings merit. While some work has been done in the shadows and in secret by a select few, without the widespread acceptance and acknowledgement of the existence of unseen entities, the type of mainstream study and research that is needed to fully understand them, what they are, what they are capable of, where they come from, what risks they pose, and so forth is not fully understood. For centuries, alien life-forms have existed among humankind and yet very little is known and understood about their actual nature and abilities. It is time for modern society to accept the reality of unseen entities and to remove the shroud of ignorance that's prevented the world from having insight into a true form of alien life that surrounds us daily, though unseen by the majority of humankind.

The Gift

Many of the individuals I've come to know who interact with unseen entities speak of their ability to see and communicate with them as having *the gift*. For some, the ability to communicate with unseen entities manifested in childhood, when the children were around individuals in their family who also had the gift and/or were participating in rituals with unseen entities. From the information I've gathered, I learned that for a separate, small minority, their ability to interact with unseen entities occurred with no outside influence; that is, no one in their family or around them had any relationship with these entities or any ability to interact with them. For others, particularly for some participants in Santeria, they had no natural ability to interact with spirits, but only after paying a *santero*—in some cases, up to US$25,000 or more, and then only through a series of initiation rituals—did they gain the ability to communicate with and see unseen entities.

One of the *curanderos* I spoke with believes these entities will interact with an individual only after that person has shown an interest in them. Whereas another *curandera* born with the gift stated that it is the entity who chooses the individual, often without the individual showing any interest and often refusing to interact with the entity for years. This *curandera* also believes that the gift is something that runs in the blood of a person and sometimes skips a generation.

Who is right? Perhaps they both are. Perhaps for some, the gift is something they are born with and an entity chooses them without any interest being shown on their part, whereas for others, once they gain knowledge, experience, and awareness of the reality of unseen entities, their personal interest brings one or more entities to them.

DNA analysis of healers born with the gift and families in which the gift has manifested through several generations is a significant research interest for me. Still, another possibility is to simply interview an unseen entity. Having a focused interview with an unseen entity to gain their input on a host of topics is a subject that fascinates me. Certain reference materials on unseen entities speak of ways in which a person can discuss any topic with an entity and gain input from them. For centuries, such knowledge has reportedly been hidden and used primarily for the personal gain of the individual possessing such knowledge. Perhaps in the future, the wisdom and insight of unseen entities will be gathered on a host of modern subjects, particularly areas of universal societal importance such as health, the environment, religious faith, and interpersonal relationships.

The ethics of the gift

One of the most fascinating aspects of the world of shamans, *paleros*, and others who communicate with unseen entities is the unique code of ethics that often guides the work they do or, in many cases, refuse to do.

There is enormous variation in the manner and the nature of the work various healers and others do with unseen entities.

Many shamans, *paleros*, *curanderos*, *santeros*, and others who work with entities exclusively for healing often hold themselves to very high moral and ethical principles. Frequently, there is a period of celibacy for *curanderos* and *santeros*, typically a year, during which the individual focuses on cultivating their spirituality and denying themselves of sexual intimacy to enhance their ability to interact with unseen entities.

There is also a common, nearly universal principle these healers have for honesty. The overwhelming majority of healers I've met do not charge for the work they do. They often view their ability to interact with unseen entities as a gift from God and, as such, freely dispense their assistance, even though they may live in poverty and survive off the generosity of their community and the individuals they serve.

For some who interact with unseen entities for monetary profit, there is considerable difference in their work and the manner in which they do it. For many healers, the thought of using their ability to injure others intentionally is repulsive. While the protection rituals healers do for others may result in injury to individuals attempting to harm a person they've protected, they see such injury as the result of the person's malevolent intent seeking to harm. Work is never done to intentionally harm except in unique instances.

For others who interact with unseen entities for profit, using their ability to harm or heal is often considered a transaction. The entity is contacted and asked what is needed to accomplish the task. The work is then completed according to the entity's direction, often with an offering or gift given to the entity in exchange for the task to be accomplished. Sometimes entities considered benevolent are also used for this work, even work that has malevolent intentions. Such activity encourages one to look at unseen entities without absolutism. Sometimes, for example, malevolent entities can be used to heal and benevolent entities to harm.

Many interactions with unseen entities involve what I would consider a transaction, that is, the entity is granted something, be it blood, animals, rum, or some other organic material, in exchange for the assistance it is being asked to provide. What is the actual benefit these entities receive from such exchanges? Does the blood somehow strengthen the entity? Spirits are considered entities without physical form, and the words of various healers suggest that their power is not without limitations. One *curandero* mentioned that the entities he interacts with often must be fed before they can be used for healing. How is it that blood and other organic materials feeds them?

By far, one of the most fascinating experiences I've observed is the *change of life* ritual. Here, the life-force of a person, usually someone elderly and infirm, is transferred to an entity in exchange for healing a person who is seriously ill. The ethics of this ritual are complex and fascinating. In some instances, the person's life-force is given willingly, making the ritual a form of assisted suicide. For others, the ritual is done without consent and can affect more than one person. At times, several individuals' life-forces are taken unwillingly. This is obviously malevolent action by both the entity and the practitioner cooperating with the entity. It raises not only ethical questions regarding the nature of unseen entities and certain practitioners but also legitimate questions about the degree of danger unseen entities pose.

Part of the impetus for scientific inquiry is to understand the nature of unseen entities more thoroughly because as much as healing and clairvoyance can be of beneficial use, the destructive power of unseen entities is something that should also be respected.

Militarization of Unseen Entities

The healing and other beneficial works performed by unseen entities are what primarily captured my interest and sparked my inquiry into the world of shamans and others who interact with what they often describe as spirits. My early exposure was to individuals who work with unseen entities for the benefit of others. As I became more informed and gathered accounts and experiences from various healers and others, I realized that other, powerfully gifted individuals interact with unseen entities for the injury of others.

These individuals' work has tremendous societal and military ramifications and applications. Despite my personal feelings and misgivings about such activities' ethics, I have salient thoughts about both the potential dangers and usefulness that unseen entities and certain individuals who work with malevolent intentions have.

What they pose to society can be harmful but can also be used in ways some may find of value.

It lacks nuance to think all who interact with unseen entities for malevolent intentions are evil individuals. Some who carry out such work consider themselves agents of justice and harm only those who have injured others seriously. Some follow the same principle as many healers by not accepting payment for their work and by not harming others who have done nothing the practitioner considers significantly evil. For others, without such ethical limitations on their work, their potential contribution to society is also worthy of consideration.

It's the nature of nearly every technological advancement to have both the potential for benefit and the potential for harm to society. It's not infrequent that some discovery's application leads primarily to the damage of humankind and to the environment. Whether or not such use is ethical will not prevent some people from using their ability for their own financial profit, regardless of the potential harm to others. We are wise to consider the potential military applications of unseen entities' powers, as it is a reality that those abilities are often applied with malevolent intentions.

Clairvoyance and the ability to directly injure individuals have obvious military application. To know an enemy's plans, their whereabouts, if someone is a spy, and what an enemy's interests might be are crucial and potential applications of interacting with unseen entities. Powerful individuals who work with spirits for malicious purposes can injure and kill from any distance for the right price. A tarot card reader, that is, one who claims to be guided by an unseen entity, describes having the ability to know any information they desire about the present through the cards' use, including the ability to find lost persons and objects. Imagine knowing where an enemy's weapons facility is located or what weapons an enemy is developing. Imagine learning these things without exposing other individuals to danger or harm and without the use of satellite or other technology. This ability is priceless and is a potential source of revolutionary military application.

Clairvoyant soldiers, bodyguards, detectives, and assassins are not a thing of imagination. The development of military technology as it relates to unseen entities, and individuals with the ability to interact with them, has obvious and tremendous value. It is likely to develop rapidly as science makes progress in understanding some individuals' ability to interact with unseen entities.

Many other potential applications have value. Frequently, the individuals doing work to harm others do so because they would otherwise live in poverty. Finding an alternative way for them to use their abilities ethically would be practical and beneficial to society.

The Dangers of Unseen Entities

Are unseen entities dangerous? Absolutely. Unseen entities are not all the same in terms of their ability or morality. This is the clear consensus of the healers and others I've interviewed who interact with them. There is also tremendous nuance when it comes to understanding unseen entities. Many individuals who work with them speak of both good and evil spirits. They also give accounts of benevolent spirits being used for malevolent actions, and malevolent spirits being used for healing and other benevolent activities. Understanding the nature of the ethics and morality of unseen entities is a complex and rich avenue for potential study.

Unseen entities are vastly superior to humans in intelligence, power, and ability and thus have the capacity for more benevolent and darker, more malevolent actions. It appears the activities of unseen entities as relating to healers and others who interact with them are commonly transactional. The entity requests something in exchange for the action or guidance they give to humans, especially for more intensive requests made, such as in cases of serious sickness or other requests. At times, however, there is no transaction for the wisdom or guidance given, particularly in respect to spontaneous moments of clairvoyance.

There are also many instances when unseen entities harass, injure, and, in some cases, possess individuals, with significantly harmful effects. Several healers I spoke with describe examples of harassment from malevolent entities leading to mental illness and even suicide. Sometimes the healers were themselves the victims of harassment and disease brought by the very entities they eventually partnered with to accomplish years of benevolent actions. The harassment and infirmities inflicted upon them were done with the intent of coercing the reluctant healer into a cooperative relationship with the entity or entities.

One *curandero* I spoke with mentioned a family that was nervous about an entity that liked to make noises in their house. The noises disturbed the parents and the children, and the father asked the *curandero* how to remove the entity from the home. To my surprise, the *curandero* said that this type of entity is not harmful but merely playful. He told the father that the entity was actually protecting his family and that they should not remove it but should learn to live with it.

According to some of the healers I spoke with, entities are like humans. They have a sense of humor and take pleasure in various activities, some harmful and some harmless. Some entities are thought to enjoy torture, blood, and sexual behavior that many would consider deviant. Some serial killers have spoken about being harassed by entities, and after hearing the reports of some of the individuals I've interviewed, I believe there is merit in some of these killers' claims.

In the section titled "The ethics of the gift," I wrote about a ritual called the change of life, one of the potential dangers involved with unseen entities. In that ritual, one or more individuals' life-force is given to an unseen entity in exchange for the entity healing another person who is gravely ill. Mattias, a *curandero* in Cuba, described an instance of a change of life ritual he did to save his mother's life. During this ritual, a spirit asked to be able to take the life of several patients at the hospital in exchange for saving the life of Mattias's mother (account contained in *The World Hidden*). It is unclear how many patients died or were injured during the course of this ritual, but the result was that his mother was healed, as were several other patients in the ICU. I think it's notable that the entity did more healing than was asked, but it is unclear the degree to which others were harmed. This ritual raises many questions about the potential dangers of unseen entities.

Of all the rituals I've seen and been told about, the change of life ritual is perhaps the one I would most like to see studied and gain more understanding about. The ethics of this ritual are, without question, troubling, but in the context of assisted suicide, there is an argument to be made about its place in ethical medical practice. Regardless of the ethics of or how we may personally feel about assisted suicide, the change of life is a ritual that takes place and reveals a lot about unseen entities' nature. Does the phenomenon of an entity absorbing another person's life-force occur in circumstances outside this ritual? If not, why not? What are the limitations unseen entities have, both ethically and physically? Are there measurable changes in blood chemistry, brain activity, and other biometrics of the individual(s) being healed and those being harmed during the ritual? Does the ritual itself reveal any other severe, hidden dangers unseen entities pose? There are numerous scientific and ethical questions that this ritual provokes.

It is not insignificant that unseen entities have coexisted with humans for millennia. Shamanic traditions in cultures worldwide suggest that benevolent spirits, as some describe them, have been sources of useful guidance, support, and acts of healing for the peoples and communities these shamans serve. The cultures that include shamanic practice are cultures where the people coexist in nature with a lifestyle that results in little to no sustained damage to the community surrounding them. The entities themselves are fed and sustained with organic material, and the healing direction unseen entities give is often rooted in the animal and plant life available to the shamans in their locale. This symbiosis leads to mutual respect for the unseen entities and the natural world, which provides the nourishment the entities request. This respect for the natural world affects the community at large and is something that modern society has significantly lost touch with, to considerably deleterious effect.

Reconciling Shamans, Spirits, and Science

Could unseen entities, commonly referred to as spirits, actually just be poorly understood alien life-forms? Are unseen entities real? Yes. But how can unseen entities be explained scientifically? The truth is that intelligent alien life-forms do exist and have been in contact with humans for thousands of years—but in a form and a manner that have been shrouded in the realm of the mysterious and the religious. Many people around the world are convinced that unseen entities are real, and there are very legitimate reasons why they do.

For years, science has ruled both my mind and my heart. As a child, I gravitated to science in school, mostly because it was one of the few things I could fully trust and rely on. Early on, I understood there were things that were definitive, certain, and undeniable in the natural world. Studying and learning about the nature of the physical and biological world provided me with a tremendous sense of fulfillment and security knowing there were things unfailingly true and real.

From the time I was five years old, I had clarity of purpose: *I knew I was going to be a doctor*. I wanted to be of service to others, and there was little that truly fascinated me the way the human body did and still does. The prestige of being a doctor also appealed to me. When a five-year-old says with confidence he's going to be a doctor, people stand up straight and look at him differently, with far more respect, and I reveled in that feeling as a child.

I dedicated myself to studying and pursuing that goal for years. Still, by the time I finished my undergraduate degree, there were deep psychological needs and questions within me that couldn't be answered and fulfilled through science alone. For many questions about life outside the classroom and the textbooks, I had no satisfying answers. And some of these are life's most important questions. Science was, and is, simply ill equipped to provide satisfactory answers to many of life's essential philosophical questions.

Science is, of course, an excellent resource for answering the questions of how and why many of the physical phenomena happening around us occur. But it's not so good at answering questions about how humankind should behave morally and ethically. Questions like, Why do we exist? Is there some purpose to life? Why does evil exist? Is it better to be noble and compassionate or to be more concerned with one's own needs? Does suffering have a purpose? What happens when we die? Why do we die? Do we live on after death? Does God exist? Does anything exist outside the universe? Such questions are just as important to understand as any scientific inquiry.

As I asked myself these and other essential questions, I began searching through various religions for answers. I returned to that empty feeling I'd had as a child as I tried but failed to understand why suffering is so prevalent and why apathy fills so much of humankind. I searched in philosophy to try to better understand the nature of humankind and the world, and yet still found my mind and heart unfulfilled and incomplete as I tried to balance faith with reason.

As I studied the Bible and the Quran and other religious and history books and philosophies, I grew determined to know if there really is a God or some form of higher power that exists outside the realm of the scientifically explainable. Despite my search, nothing ever led me to any concrete evidence that gave me satisfactory answers. Religion is just, simply, largely based on faith.

That period of philosophical and religious study proved invaluable once I began considering the nature of unseen entities. I read many accounts that I found intriguing of spirits from religious texts and books on the occult. Do beings that have no permanent physical form really exist? Is it possible that they have the enormous intelligence and power described and have interacted with humankind for thousands of years? These questions seemed to belong more appropriately in the realm of fantasy, science fiction, or religious philosophy. They didn't seem like something a scientifically oriented person should give any significant consideration to. For years, I dismissed these accounts as insignificant until my own transformative experience.

I had an epiphany and asked myself a fundamental question I'd never considered before my experience: Why does nearly every culture that has ever existed have some form of medicine man or shaman who claims to interact with spirits for the benefit and welfare of their community? As I considered this question, I began reading more about shamans and various cultures that were, and perhaps still are, separated by both vast geographic distances and at times centuries of time in origin. I found similarities in the methods, characteristics, attitudes, and practices of various shamans despite a significant degree of cultural and geographic difference between them. It's obvious such similarities are beyond mere coincidence. Either the accounts of shamans and the unique abilities ascribed to them are accurate, or nearly every people who ever lived suffers from the exact same delusion.

Though not always the case, in nearly every culture in which a form of shamanism exists, there are patterns of characteristics, practices, and experiences commonly displayed among those claiming to communicate with unseen entities for the healing of their community. Here are some of the most frequent similarities.

1. *The society.* Shamans live in a society or community that accepts that unseen entities are real and acknowledges that the shaman is endowed with

special abilities as a result of their ability to communicate and interact with these entities.

2. *The person.* The shaman frequently has an eccentric personality and, at times, though not necessarily, unique physical characteristics like an abnormal number of teeth or digits or some other atypical bone structure, but this is not universal. There is nearly always a hyperdeveloped sense of intuition and sensitivity demonstrated by the shaman that helps to distinguish them from others in the community.

3. *The medium.* Shamans have the ability to communicate with unseen entities, oftentimes directly but also through the ability to enter trance-like states where they are able to interact with the entities and/or channel the entities through their body. When going into these trance-like states, as well as while healing and performing other rituals, the shaman often uses the assistance of certain objects such as bull horns, feathers, a large sea turtle shell, drums, a glass of water, headgear, a gown, a metal rattler, a mirror, a staff, and/or other objects. The specific shapes and properties of these instruments are useful in various ways, depending on the specific form of shamanism practiced.

4. *The guide.* Shamans are often assisted by an active guardian entity or group of unseen entities that have distinct personalities and interests.

5. *Resistance to the call.* The unique and powerful abilities of the shaman are believed to result from a choice made by one or more unseen entities. The one who is chosen—often while still a child or adolescent—may resist this calling, sometimes for years. Harassment and/or torture from the entity, appearing in the form of physical or mental illness inflicted on the person or someone they love, breaks the resistance to the call of the entity or entities and the shaman accepts the calling to work with the entity or entities as a healer.

As mentioned previously, the impetus for writing *The World Hidden* initially came from my own experience with a shaman, a *santera* who lives in a small province near where my wife grew up in Cuba. For some, referring to a *santera* as a shaman is a significant inaccuracy. Still, I chose to use the term *shaman* in the book to refer to any healer who interacts with spirits for the sake of healing. I did this as a way to help the reader unacquainted with the world of spirit-guided healers. However, I am aware of the significant cultural and religious belief system differences between various individuals who interact with unseen entities.

My experience led me to seek to understand more about Santeria, the religion that the *santera* practices. While Santeria has its own rules and rituals and its practice is predominantly rooted in the Caribbean, parts of the US, and South America, I immediately felt that there would likely be others from various other cultures around the world who work with spirits. I'd noticed this studying religious texts from various cultures. I viewed diverse global shamanic figures back then and still, to this day, as various branches on a tree that share a common root, despite there being differences in the various beliefs and rituals individual shamans use.

I left Cuba convinced that unseen entities are real. Though certain unseen entities, often referred to as spirits, are real, I still had no clue how to explain these entities scientifically. Everything I learned about unseen entities before my own experience came from religious and occult texts and movies. I had no other point of reference to unseen entities beyond the influence of my personal religious and cultural influences. But are religious and occult books really the definitive experts on spirits? To try to scientifically and objectively characterize the nature of unseen entities based on guidance from religious and occult texts, particularly, some of which within themselves claim to be "inspired by God," creates a number of challenges—not the least of which is, Does God actually exist? That is far too lofty of a question for me to answer, for it's certainly scientifically unprovable at present, so I decided to focus on something that seemed more accessible and reasonable. Could it be that what people commonly refer to as spirits, entities often viewed only through the lens of religious mysticism, fear, and faith, are very real intelligent life-forms existing outside a physical form we can easily comprehend? Could it be that Indigenous and other peoples from times long past, including writers of the Bible and religious texts, were describing these entities in religious terms because they had no other lexicon or perspective with which to explain them?

As I began considering the nature of unseen entities, I reluctantly accepted that the primary sources of literature on the subject were religious books and books on the occult, neither of which would be met with a significant amount of credulity or reliability from scientifically oriented minds. Yet, regardless of that ambiguity, I decided to consider the texts anyway and see if perhaps there were nuggets of truth in them that would help me in my consideration of the possibility that unseen entities are real. After all, at least religion and the occult are aware that unseen entities exist, which shows that, in some respect, they are much further along than our current scientific understanding, which doesn't even acknowledge the reality of these entities at all.

For the vast majority of cultures in nearly every epoch of human existence, shamans have existed in various forms. In truth, there are significant differences in the belief systems of individuals who interact with spirits. For example, cultural heritage and religious framework affect the way a *santera* uses rituals to interact with unseen entities in comparison with, say, how a Buddhist shaman interacts and works with them. The individual spirits a *santera* may interact with are frequently specific orishas, that is, beings with specific likes, interests, and personalities. The Buddhist shaman, however, may not work with a widely known or popular spirit but rather one familiar to him only. Nevertheless, despite ritualistic and belief system differences, there are many similarities that can be found in the manner and practice these and other individuals who interact with unseen entities share.

For example, while a shaman from a Native American tribe may use feathers or animal skulls or bones as tools in rituals and interactions with unseen entities, a *santera* may use a large sea turtle shell or dolls as a way to connect with them. While the specific objects may be different, there is still a common theme in that similar objects are used; at times, the same objects are used. Shamans also share other tools, like the use of chicken, pigeon, goat, or other animal blood, in rituals.

Another example of similarity in rituals is found in the use of local herbs, such as peyote in North America and ayahuasca in South America. While very different pharmacologically, both of these substances fling their users into hallucinogenic states with considerably more lucidity than, say, PCP or ketamine. Were the shamans using these substances and providing them to their communities in rituals, merely trying to get people stoned so that they could deceive them into believing spirits are real? That certainly is one potential point of view. But what do people who've actually participated in such rituals describe? Certainly not some kind of hallucinogenic-driven party; rather, they describe deeply spiritually enriching experiences that provide some sort of timely direction and guidance needed in their lives.

Convinced as I was by the authenticity of the *santera* I met and by the awareness that unseen entities do exist, a fire lit inside me to try to understand the nature of them. I knew that there would be other shamans and people who have consulted with them that likely have experiences similar to my own.

My wife grew up in Cuba, an island nation where Santeria is practiced heavily. Santeria is a religion where the existence of unseen entities is not only openly acknowledged but also plays a central role in the practice of the religion. Rituals are used to gain favor with specific entities, and priests of the faith claim to both see and communicate with those entities. Acts of healing, divination, and guidance are often provided for the assistance of those in the presence of the priest interacting with an entity or entities.

I began talking with my in-laws and friends in Cuba about my experience. In return, they shared some of their own experiences with *curanderos* and *santeros* and, in some cases, *paleros*. I came to see that there was a wealth of opportunities for me to receive answers to some of my many questions within the community of friends and family I have in Cuba. So, I sat down with them. Their accounts and points of view fascinated me. I became convinced that their accounts of unseen entities and shamans had tremendous value to the public, largely unfamiliar with these entities and what they can do. I then met other individuals and listened to their experiences. I met with reputable *santeros*, *curanderas*, shamans, *paleros*, and others who interact with unseen entities around the whole of Cuba and the United States, particularly the US Southwest. We, too, sat down and talked. I grew even more convinced that their experiences had merit and should be considered by the scientific community and the public in general.

I live and work in New Mexico near several Native American reservations. There are many practicing shamans, *curanderos*, and others who interact with unseen entities in the area. I've been fortunate enough to associate with some of these individuals and enjoy their cooperation in sharing their experiences with unseen entities. The accounts detailed in *The World Hidden* describe several common abilities that cultures with shamans throughout Earth describe these unique individuals as manifesting as a result of their relationship with unseen entities.

Clairvoyance

Clairvoyance is the ability to perceive events in the future or things beyond normal sensory contact and to see and/or hear unseen entities around them. Nearly every known culture that has, or had, shamans reports that shamans have this ability. For some shamans, clairvoyance comes from direct communication with unseen entities. For others, it occurs through the recognition of signs, physical phenomena occurring around them, or some organized form of divination, such as *caracoles*, which indicates what a future outcome may be or will be. Clairvoyance occurs through varied and diverse means, whether through direct communication from unseen entities through various means, and through what can be accurately described as a hyperdeveloped sense of intuition. There is often great variability in the way different individuals who interact with unseen entities perceive clairvoyance. Yet in each individual's case, that experience is often associated with a definitive clarity that allows that particular individual to understand the degree to which some event they are physically removed from is happening or will happen and/or its degree of likelihood to occur in the future.

Healing

Healing is the ability to gain knowledge of physical and psychological ailments through the guidance of unseen entities. Unseen entities often direct the solution to the ailment through various means. At times, herbal preparations are used. At other times, prayer, blowing smoke, and energy cleansing with leaves and/or an egg are all that are needed. In cases of more serious health problems, animal sacrifices, including the use of blood, are required to accomplish healing.

Medium

While not all shamans and others who interact with unseen entities channel them into their bodies, many do. This ability to channel these entities often results in the manifestation of various abilities from these entities, or a more potent expression of them. For example, acts of healing, clairvoyance, and/or a clearer perception of the problem(s) affecting the individual seeking assistance occur more intensely as an unseen entity is channeled through the shaman. Terms such as *in trance* and *in transition* are often used to describe periods when a shaman or other person is actively channeling an unseen entity through them.

For most mediums, there is a gradual process that occurs over time that helps them to understand what they can do with these entities. Frequently, a person begins to see unseen entities in childhood (i.e., an imaginary friend), and the entity or entities engage in activities to communicate with the person, often through direct conversation and repeatedly visiting or touching the person, often while they are resting or attempting to sleep. Often there is an initial reluctance on the part of the person with the ability to become a medium and accept that unseen entities are real. There is frequently a period of refusal on the part of the individual, during which the entity, often through coercion, forces the person to accept what they are and to work with the ability they have been given to be of service to others. The coercion is often achieved through the entities making either the shaman or someone close to them ill and then providing the healing to the one affected, as the shaman accepts their gift and follows the direction of the entity. Many of the accounts in *The World Hidden* illustrate this pattern.

The ability to affect outcomes

A fascinating ability both shamans and others manifest through their interactions with unseen entities is the ability to direct favorable and/or harmful outcomes to others. Shamans generally work to accomplish positive outcomes. At times, the effects of protection rituals can harm someone seeking to harm an individual or individuals the shaman has performed a protection ritual with. In so protecting an individual or individuals through their work with unseen entities, the harmful outcome may return to the individual seeking to harm someone protected. The shaman often views this consequence as the natural outcome of the malevolent intent of the person seeking to do evil.

This ability to affect outcomes commonly pertains to health, but also, through their interactions with unseen entities, shamans and others can affect legal outcomes or encourage individuals to take actions they might otherwise resist or reject in matters of love, business, politics, and many other arenas. Frequently, a person working with unseen entities to bring harm can similarly accomplish the desired outcomes through direction from unseen entities to affect the person targeted, or through violence the entities can encourage in others, or through accidents, theft, and other actions. The manner and mechanisms of such unique activity are fascinating and engender a tremendous number of legitimate questions regarding how unseen entities can affect these actions. Many accounts reference these realities.

After considering such accounts and personally experiencing similar things in my dealings with shamans, I was moved by the depth of humility that I've found in these individuals. Even the noblest person can be easily tempted to develop feelings of superiority and arrogance when endowed with common human advantages like wealth, intelligence, and prestige. Yet the overwhelming sense of humility and modesty in shamans, who are often endowed with the ability to achieve incredible feats of healing, clairvoyance, and other deeds, begs the question, Why such modesty?

Such modesty has two obvious and immediate causes, which are readily discernible. First, the individuals gifted to interact with unseen entities are often those with an appetite for what is often considered spiritual things. The shamans, *curanderos*, *santeros*, and other practitioners I know who use their work with unseen entities for benevolent actions pride themselves on having character of the highest caliber. They often view such character as inherent to their ability to maintain favor with the unseen entities that endow them with the extraordinary gifts they manifest as well as an awareness that their abilities are exclusively the result of sources outside themselves. They understand that such works have nothing at all to do with any unique ability within themselves but is the fruitage of their partnership with and relationship to the unseen entities whose strength and power flow through them.

Second, like those who use their relationships with unseen entities for less selfless reasons, the modest ones understand, often from personal experience or through the direct observation of others, that when the unseen entities are not honored or given the respect and recognition they merit, the flow of power from them is stopped. These entities can also be angered into harming a person who once was allowed to channel the entity's ability if that person fails to honor some promise or conduct themselves in ways the entity feels are appropriate.

I realize that for many of the individuals who work with unseen entities, the use of the term *alien life-forms* as it relates to what are commonly known as spirits is not ideal. Opinions vary among shamans and others concerning the origin and nature of the entities they interact with. For some, spirits represent nothing less than deities and are, in fact, prayed to and worshiped, whereas, for others, spirits are less than gods and more like benevolent helpers. Still, others see spirits as a form of life that lives on after our physical body has died. Beyond these few opinions, there are still many other beliefs. The spirits of the ancestors, for example, is a common belief among many African peoples and various Indigenous cultures around the world. It is not due to a lack of respect that I refer to unseen entities as alien life-forms; it is with an awareness that unseen entities are, in fact, intelligent, nonhuman entities whose origin and exact nature remain disputed even among shamans. This lack of congruence in thought among those who interact with unseen entities is another reason why further study is necessary.

So, I invite you into the realm of some of the world's true alien life-forms. Not only are unseen entities real, but they are enormously intelligent beings in possession of powerful abilities and understanding far beyond humankind. For centuries, their presence has been hidden in the realm of the mysterious and religious, hiding in plain sight a truth so obvious and yet one only a few, relatively select, gifted individuals among humankind have come to know intimately. Alien beings have watched, communicated with, and interacted with humankind for thousands of years.

Chapter Two
To Dust All Return

> [19] Surely the fate of human beings is like that of the animals: the same fate awaits them both: as one dies, so dies the other. All have the same breath; humans have no advantage over animals. Everything is meaningless. [20] All go to the same place; all come from dust, and to dust all return.
> —Ecclesiastes 3:19–20 (*New International Version**)

> [18] I hated all the things I had toiled for under the sun, because I must leave them to the one who comes after me. [19] And who knows whether that person will be wise or foolish? Yet they will have control over all the fruit of my toil into which I have poured my effort and skill under the sun. This, too, is meaningless.
> —Ecclesiastes 2:18–19

Nobody talks shit like a Cuban, except maybe Black dudes and cowboys. I've known my fair share of all three through my travels as a writer and photojournalist, and especially during my time as a student at Morehouse. The cowboys might surprise you—trust me, cowboys are some of the best shit talkers. Maybe it's from poverty, not that poverty is a requisite or a universal condition for them, or the monotony of long days of hard work, but invariably, people who know lives of struggle often have a great sense of humor, for surviving routine hardship, to me, seems to mandate such.

---

* Scripture quotations are taken from the *New International Version* (NIV), unless otherwise indicated.

My wife and I are opposites in just about every way possible, except for our sense of humor and our love of salsa dancing and bachata. It was her freedom with words, a boldness to say how she feels with little regard for the way others feel about it, that first captured my attention and my respect. She's usually reserved in public, especially with strangers, but when you know her, or if you offend her, her guard drops, and she's not afraid to let you know and command her respect. Most of my happiest moments with her are just our long back porch conversations, talking shit, and enjoying a glass a wine and a smoke.

July 11, 2021, will be a day long remembered in Cuba's history, and I suspect history will remember it as the first day of Cuba's new revolution for freedom. When President Trump came to office, he reversed Obama's policies on pretty much everything, including Cuba. It was a course I personally resented, mostly because it would make things harder for me and my family to visit my wife's large family still residing on the island and to send them money. I'd grown fond of Cuba and its people from our visits there and made a large collection of friends on the island. I became resentful of the overwhelming poverty that engulfs the island. Cuba is a jewel, and its people are precious to me for a number of reasons—not the least of which is the potential I see in an industrious people denied freedom and opportunity.

When I initially went to Cuba, the first few times especially, I hated the poverty I saw there. There's an inherent injustice in poverty that beckons one to seek out its source and give thoughtful consideration to finding its remedy. It was not my own welfare that bid my thoughts in such direction, but a wishing, a hoping, that my friends and family on the island might know a life worthy of their benevolent character.

There seems no greater measure of the failure of human civilization and society than that a hardworking and honest person must endure poverty for the sad misfortune of nothing more than the circumstances of where they are born. This sad actuality, more than any other reality, shows both the failure of human civilization and the need for *one global government*. Poverty exists solely because of greed. There is no reason any person should starve or be forced to live in squalor because of where they are born. For this, and society's many other failures, the world will not continue this way. Both nature and the spirit world that remains largely hidden move toward balance, and the harmony that such balance mandates means sure condemnation for the world as we now know it.

It is a failure of gross proportions for humanity to think of Earth as our puppet; it is not, it is our *mother*. Our survival as a species rests upon indefinitely obtaining the resources we need from this planet. Earth is not our slave but our provider, and as we have grown beyond the proper respect for our provider as a species, so too, within the framework of its existence, are built the mechanisms needed to bring balance to this disharmony.

"In fact, unless that time of calamity is shortened, not a single person will survive. But it will be shortened for the sake of God's chosen ones." So reads Matthew 24:22. I am no scholar of the Bible or its many prophecies, but I see the parallel of this scripture and the actuality of what is taking place on Earth and I feel its truth. When governments fail humanity, when nature's repeated calls for humanity to stop overtaking are ignored, when interpersonal relationships fail to produce the type of society required for our species to indefinitely survive, I cannot fail to believe that Earth will not act in its own defense. Humanity has failed as its defender and protector, and so it must now protect itself.

The changes we see occurring globally to Earth's climate, the rising sea levels, the warming oceans, the appearance and proliferation of novel infectious diseases, and so on, speak of a cycle of extinction within which Earth now firmly resides. It is a fear-inspiring actuality that, in itself, guarantees the future survival of both humans and other species, for from death comes life. The cycle our planet now descends into will in time result in the salvation of humanity and many other species, but not without cataclysmic suffering and loss of human life before such. As Earth becomes more unstable, it becomes more uninhabitable and less capable of providing food for the sustenance of humanity, and the eventual death of large segments of the human populace will in time bring the restoration of balance between Earth and its dominant species, to the benefit of every other species on this planet.

The spirit world relies on and cares much for Earth, as it is, and Earth has been the sustenance for many of these beings and the focus of their care. They are not without skin in the game, as it were, and our species' survival is inherent in the spirit world's quest for balance among humanity, Earth, and itself. Indigenous prophecies speak greatly of the return of this balance, and the sobering words of Matthew 24:22 parallel many Indigenous philosophies that speak about a return of balance between Earth and humanity. There are a great many sobering Indigenous prophecies, which are discussed in a later chapter, and they simply cannot be ignored in this time of unprecedented global catastrophe and pandemic.

Our Shared Ancestral Bond

It is not a trivial accomplishment, nor one that happens naturally, to deeply appreciate that the most valuable possession every living person owns we owe entirely to individuals other than ourselves. Every person who has ever lived, the entirety of our being genetically, our unique and inherent natural talents and gifts, is but the fulness of our ancestral identity. The whole of our physical being is solely the summation of generations of our ancestors who came before us. The homage Indigenous peoples—African, Native American, Aboriginal, Inuit peoples, and others—pay to their ancestors is rooted in recognition of this profound yet indelible truth. Taken to the inception of our species' origin, every human shares the same originality, the very same progenitors. It is with this recognition of our indelible dependence upon this shared ancestral bond that I speak now on superiority.

When we look at our world, we see a planet and its dominant species, fractured and disconnected. This lack of harmony and balance between the natural world and humanity starts first at the microcosm of the family units and the local communities that make up our vast civilization. Our environment, treated without the appreciation it merits, spirals into greater instability by the day. The presentation and proliferation of unique and novel diseases like COVID-19, a virus some suggests came from either a lab or a species of bats humans typically have no close interaction with, reflects the reality that humanity has encroached beyond an appropriate position of harmony with the natural world that sustains our existence. This imbalance has catastrophic consequences, not just for our species alone but for all life on Earth.

Oppression and abuses of all manner occur in every part of the world. Homelessness, poverty, starvation, and war are all conditions of choice, albeit most often not the choice of the individuals most afflicted by them directly but the decisions of a select few among humankind, *The Masters of Mankind*, as named by Adam Smith. These masters' callousness allows the suffering we see present throughout the world to proliferate and persist.

Earth has enough raw materials to provide adequate housing for all, enough food that no person should starve, enough space that no persons should be impelled to squabble and be ensnared in cycles of violence to obtain the necessities life requires. Such suffering and inequality in the face of abundant resources illustrates the depth of toxicity that permeates human civilization. But what is the root of such suffering?

One of the greatest tragedies affecting humanity today has afflicted it throughout its history. The ignorance of a subtle yet powerful illusion plays out daily around the globe, though hidden from the consciousness of the vast majority of people. The drive for individual superiority is an impetus rooted in fallacy. It internally motivates the masters of mankind, individuals who dominate the political, industrial, and commercial realms of society, albeit most often subconsciously. The drive to be greater than others is an inherently self-destructive, community-destructive, and environmentally destructive desire. It has obvious catastrophic consequences for interpersonal relationships, personal psyche, and the greater macrocosms of society. The greed and callousness of a few make genuine happiness impossible for the overwhelming masses of humanity. The fruitage of this illusion's ferocity is reflected throughout humanity's history of innumerable conflicts between various nations and peoples.

The root of the illusion of superiority's destructiveness survives on a fallacy, on a personal foundational belief that can only blossom into actions leading to suffering and harm. This foundational fallacy is that an individual's life and value can somehow be greater than another person's, that a person's inherent value can be elevated above another's simply by what one possesses, acquires, or achieves. This illusion is the root of the great majority of our civilization's problems and ills, and the reason genuine happiness eludes most people.

Dehumanizing and Self-Destructive Behavior

The illusion of superiority leads to an indolent but subtly destructive exercise: comparing oneself with others. This basic learned human behavior leads to tremendous suffering, both to oneself and to others. Comparing oneself to others can lead to only two possible outcomes: a person will consider themself either superior or inferior to the individual in comparison. Both outcomes inevitably lead to unhappiness and suffering.

    The individual who concludes they are superior to others devalues the person or persons considered inferior. This devaluing of individuals' lives ultimately results in a loss of interpersonal solidarity and connectivity, which results in powerful dehumanizing effects. The apex of such thinking, extended in its extreme, allows for the pathologic psyche of usurping another person's liberties, possessions, and potentially even another's right to life, as exemplified in a willingness to go to war or commit murder. To consider the value of another person's life inferior enables one to exploit and otherwise take advantage of others in the most grotesque and egregious of ways, as exemplified repeatedly throughout humanity's history. Even the most heinous of evils finds root in the fundamentally erroneous belief of one's superiority over others. The Nazi Holocaust; African, Caribbean, and South and Central American colonization; American slavery and the Jim Crow period after; the Native American genocide; and many other atrocities all find their history rooted in this most basic illusion: the belief of one individual or group in their superiority over others.

    The comparison of oneself with others also has the potential to cause enormous self-inflicted suffering and self-destructive behavior. If an individual finds themself to be inferior to others upon comparison, this will inevitably lead to feelings of inadequacy, depression, self-loathing, and, ultimately, self-destructive behaviors. This is one of the most significant reasons depression so commonly afflicts humanity, for our civilization is deeply rooted in this fallacy. Social media only reinforces and facilitates comparison with others, super-encouraging this behavior. A person who finds themself inferior to others can easily fall victim to feelings of worthlessness. These feelings often lead to further cycles of thoughts and actions that contribute to feelings of inadequacy, social isolation, physical inactivity, poor nutrition, lack of goal seeking, drug addiction, and so on.

There is one exception, however: when comparison with another individual motivates a person who judges themself inferior to improve their choices and lot in life. This is the only healthy outcome of comparing and is often the fruitage of good leadership and companionship with others whose lives provide examples worthy of imitation. In this sense, when feelings of envy, jealousy, and self-loathing are avoided, comparison can positively yield personal growth and development.

All are the same in what we have been given... life.

The value of an individual's life cannot be inferior or superior to others in inherent value. The root of this lies in a fundamental truth: *Life in itself has equivalent value.* But what does that really mean? Regardless of what station one is born into materially, or what one acquires throughout life, be it education, wealth, fame, beauty, or any other superficial measure of personal value, every individual has but one destiny: We all grow older and die, albeit some more rapidly than others. Our shared destiny, that is, death, ensures that whatever one gains in life is temporary, leading to no lasting or significant change in our individual mortality. Any advantage one gains, be it measured materially, athletically, intellectually, socially, or otherwise, is fleeting, limited by our mortality. *From dust we are, to dust we return.*

But our equivalency is rooted in more than our shared mortality. We share a parallel journey, with similar joys, struggles, desires, and goals, and these universal affections bind all of humanity. The desire for happiness, purpose, romantic love, belonging, friendship, and personal worth is universal to the human condition and not guaranteed by the possession of any amount of wealth or any other acquisitions, skills, or natural ability one may gain in life. This reality highlights another truth: True happiness is not rooted in the external or one's achievements or possessions. It comes from internal values and connectivity, the love we feel for ourselves, other beings, and the natural world, and the benevolent actions that love impels within us. Material things, beauty, educational accomplishments, and so on all lose their value as death nears.

Individuals whose happiness is rooted in external things often grow bitter in old age as they confront their mortality and the temporary value of all things physical and material. Old age is a time when a person can know great happiness and joy if life is viewed from the right perspective and lived properly. The youth can benefit significantly from the experience and understanding of those with the possession of the wealth of longevity and the wisdom it often provides. I've known many a precious mentor whose hair was gray and white and whose pace was slow and deliberate, and the value of their time, wisdom, and experience shared with me cannot be quantified by any measure.

> All share a common destiny—the righteous and the wicked, the good and the bad, the clean and the unclean, those who offer sacrifices and those who do not. As it is with the good, so with the sinful; as it is with those who take oaths, so with those who are afraid to take them.
>
> —Ecclesiastes 9:2

Ecclesiastes is a less popular Bible book, but it expounds on life's vanities in complex and nuanced ways.

> [1] I said to myself, "Come on, let's try pleasure. Let's look for the 'good things' in life." But I found that this, too, was meaningless. [2] So I said, "Laughter is silly. What good does it do to seek pleasure?" [3] After much thought, I decided to cheer myself with wine. And while still seeking wisdom, I clutched at foolishness. In this way, I tried to experience the only happiness most people find during their brief life in this world.
>
> [4] I also tried to find meaning by building huge homes for myself and by planting beautiful vineyards. [5] I made gardens and parks, filling them with all kinds of fruit trees. [6] I built reservoirs to collect the water to irrigate my many flourishing groves. [7] I bought slaves, both men and women, and others were born into my household. I also owned large herds and flocks, more than any of the

kings who had lived in Jerusalem before me. ⁸ I collected great sums of silver and gold, the treasure of many kings and provinces. I hired wonderful singers, both men and women, and had many beautiful concubines. I had everything a man could desire! ⁹ So I became greater than all who had lived in Jerusalem before me, and my wisdom never failed me.

¹⁰ Anything I wanted, I would take. I denied myself no pleasure. I even found great pleasure in hard work, a reward for all my labors. ¹¹ But as I looked at everything I had worked so hard to accomplish, it was all so meaningless—like chasing the wind. There was nothing really worthwhile anywhere.

The Wise and the Foolish

¹² So I decided to compare wisdom with foolishness and madness (for who can do this better than I, the king?). ¹³ I thought, "Wisdom is better than foolishness, just as light is better than darkness. ¹⁴ For the wise can see where they are going, but fools walk in the dark." Yet I saw that the wise and the foolish share the same fate. Both will die.

¹⁵ So I said to myself, "Since I will end up the same as the fool, what's the value of all my wisdom? This is all so meaningless!" ¹⁶ For the wise and the foolish both die. The wise will not be remembered any longer than the fool. In the days to come, both will be forgotten.

¹⁷ So I came to hate life because everything done here under the sun is so troubling. Everything is meaningless—like chasing the wind.

The Futility of Work

[18] I came to hate all my hard work here on Earth, for I must leave to others everything I have earned. [19] And who can tell whether my successors will be wise or foolish? Yet they will control everything I have gained by my skill and hard work under the sun. How meaningless! [20] So I gave up in despair, questioning the value of all my hard work in this world. [21] Some people work wisely with knowledge and skill, then must leave the fruit of their efforts to someone who hasn't worked for it. This, too, is meaningless, a great tragedy. [22] So what do people get in this life for all their hard work and anxiety? [23] Their days of labor are filled with pain and grief; even at night their minds cannot rest. It is all meaningless.

[24] So I decided there is nothing better than to enjoy food and drink and to find satisfaction in work. Then I realized that these pleasures are from the hand of God. [25] For who can eat or enjoy anything apart from him? [26] God gives wisdom, knowledge, and joy to those who please him. But if a sinner becomes wealthy, God takes the wealth away and gives it to those who please him. This, too, is meaningless—like chasing the wind.
—Ecclesiastes 2:1–26

King Solomon was an extraordinarily wealthy man, with an annual salary in the hundreds of millions of US dollars, and primarily with gold, not some artificial invisible currency according to both Bible and historical accounts. He owned several large homes, a palace, gardens, parks, vineyards, orchards, and livestock and had hundreds of wives and concubines. He was a brilliant philosopher and city planner. He oversaw the construction of many public works in Jerusalem, including the construction of Jerusalem's first temple and the enlargement of the city's massive double walls. By the typical standard of success adopted by the world, he certainly was successful.

> [22] Solomon's daily provisions were thirty cors of the finest flour and sixty cors of meal, [23] ten head of stall-fed cattle, twenty of pasture-fed cattle and a hundred sheep and goats, as well as deer, gazelles, roebucks and choice fowl. [24] For he ruled over all the kingdoms west of the Euphrates River, from Tiphsah to Gaza, and had peace on all sides. [25] During Solomon's lifetime Judah and Israel, from Dan to Beersheba, lived in safety, everyone under their own vine and under their own fig tree. [26] Solomon had four thousand stalls for chariot horses, and twelve thousand horses.
>
> —1 Kings 4:22–26

> [3] He had seven hundred wives of royal birth and three hundred concubines, and his wives led him astray.
>
> —1 Kings 11:3

Solomon's life exemplified the pinnacle of luxury. With all that he acquired, his wisdom, the vastness of his riches, and the diversity and intensity of a sex life offered by a thousand women of diverse ethnic and cultural heritage, Solomon knew a life full of honor, fame, pleasure, and splendor. Still, he realized he was ensnared in the same great tragedy that binds all of humanity. He understood the actuality of his mortality, and that understanding proved vexatious for Solomon as he comprehended the inherent equality of his mortality:

> [18] I came to hate all my hard work here on Earth, for I must leave to others everything I have earned. [19] And

> who can tell whether my successors will be wise or foolish? Yet they will control everything I have gained by my skill and hard work under the sun. How meaningless! [20] So I gave up in despair, questioning the value of all my hard work in this world.
>
> —Ecclesiastes 2:18–20

Depending on the translation of the Bible used, the final word in Ecclesiastes 2:19 varies slightly. In the New International Version (NIV), New Living Translation, it is "meaningless." In the King James Version (KJV) and English Standard Version (ESV), it is "vanity." In the Holman Christian Standard Bible (HCSB), it is "futile." This idea is preserved regardless of the translation.

For Solomon, his mortality placed him on equal footing with his fellow man, regardless of his accomplishments, talents, social status, fame, and wealth. He concluded that everything was vanity, an illusion, like chasing after the wind. He reduced what he considered significant in life to a couple of simple statements.

> [24] So I decided there is nothing better than to enjoy food and drink and to find satisfaction in work. Then I realized that these pleasures are from the hand of God. [25] For who can eat or enjoy anything apart from him? [26] God gives wisdom, knowledge, and joy to those who please him.
>
> —Ecclesiastes 2:24–26

> The end of the matter; all has been heard. Fear God and keep his commandments, for this is the whole duty of man.
>
> —Ecclesiastes 12:13

To Enjoy the Pleasures of Life and Work

Solomon considered finding joy in the pleasures of life and satisfaction in one's work the worthiest pursuits in life. His viewpoint diverged from that of the Epicureans, who valued pleasure above all, albeit in subtle and nuanced ways often not highlighted or widely understood; for example, mental pleasures were considered the highest form of pleasure for Epicureans, beyond the often-quoted mantra "eat drink and be merry, for tomorrow we're dying." For Solomon, it was a balanced pleasure seeking, meant to be enjoyed within the confines of other prohibitions within the Mosaic Law, such as drunkenness and promiscuity, since Solomon was also a Jew. No mandate regarding the service to humanity, community, or the natural world was included in Solomon's words in Ecclesiastes, though in other collections of his writings, such activity is mentioned. For example, Proverbs 11:25, and other scriptures encourage generosity. Proverbs 12:10 speaks of caring for one's domestic animals. Still, there is no mention of nature, or preserving the natural world, concepts that likely had no reference in his time and certainly were not ideas integral to Jewish life and practice.

Religion's Effort to Make Life Less Futile

The Israelites' peroration to remain committed in their worship of God is a view worthy of considering, not for Christians and Jews alone, but Muslims and believers in reincarnation as well. For any whose religious beliefs incline them to believe their actions in this life affect their future life prospects after death, pondering on how a person should treat others is not a vain pursuit at all. Such a view is of preeminent importance; for some, the potential reward of everlasting life or an improved station in their next life is worth any sacrifices an individual might make.

Consider some Hindu text regarding reincarnation, which predates Buddhist thought in the ninth through sixth centuries BCE. **Ātman** (/ˈɑːtmən/; Sanskrit: आत्मन्) is a Sanskrit word that means inner self, spirit, or soul. The text below explicitly mentions reincarnation in the classic sense, with an individual Jiva moving from one body to the next.

> स यत्रायमात्माबल्यं न्येत्य संमोहमिव न्येति .... आत्मा निष्क्रामति .... तमुत्क्रामन्तं प्राणोऽनूत्क्रामति.... स विज्ञानो भवति सविज्ञानमेवान्वावक्रामति । तं विद्याकर्मणी समन्वारभेते पूर्वप्रज्ञा च । तद्यथा तृणजलायुका तृणस्यान्तं गत्वा अन्यमाक्रममाक्रम्यात्मानमुपसंहरति एवमेवायमात्मेदं शरीरं निहत्याविद्यां गमयित्वा अन्यमाक्रममाक्रम्य आत्मानमुपसंहरति

[When the body becomes weak and goes into oblivion as it were, the Atman departs, and following it, the vital breath departs ... he becomes a pure consciousness, and with this consciousness, he proceeds. His past learning and deeds as well subtle memory accompany him. Just as a worm upon reaching the tip of a blade of grass, reaches out towards another blade of grass by way of support, so also does this Jiva end this body, becomes imperceptible, and then obtains another body by way of support, and pulls itself together.]

तद्यथा पेशस्कारी पेशसो मात्रामपादायान्यन्नवतरं कल्याणतरं रूपं तनुत एवमेवायमात्मा इदं शरीरं निहत्याविद्यां गमयित्वान्यन्नवतरं कल्याणतरं रूपं कुरुते पित्र्यं वा गान्धर्वं वा दैवं वा प्राजापत्यं वा ब्राह्मं वा अन्येषां वा भूतानाम् । यथाचारी यथाकारी तथा भवति साधुकारी साधुर्भवति पापकारी पापो भवति पुण्यः पुण्येन कर्मणा भवति पापः पापेन । अथो खल्वाहुः काममय एवायं पुरुष इति स यथाकामो भवति तत्क्रतुर्भवति यत्क्रतुर्भवति तत्कर्म कुरुते यत्कर्म कुरुते तदभिसंपद्यते ।

[Just as am embroiderer takes out a piece of embroidered cloth and creates a newer, more beautiful form, so also the Jivatman having ended this body becomes imperceptible and then creates for itself a newer, more beautiful body—be it of the manes, Gandharva, Deva, Prajapati, Brahma, or other creatures. As he behaves, as he acts, so he becomes—good behavior makes him good, bad behavior makes him bad. He becomes better by better deeds and worse by worse deeds. Therefore they say, "man is a sum of his desires," because as he desires, so are his intentions; as are his intentions, so are his deeds. His deeds are what make him what he is.]

—Brihadaranyaka Upanishad 4.4

The idea of a person's actions affecting their future life prospects is the preeminent concept in many religions. Variations in thought on reincarnation are present in Buddhist, Hindu, and other religious traditions. Regardless of a person's personal feelings about various religious convictions, there are a great many times when those convictions compel moral and benevolent conduct such as generosity, honesty, self-control, marital fidelity, and a host of other principled behaviors among sincere religious devotees. In as much as religion is often maligned for the conflict it engenders—

and rightly so considering the history of religion to tolerate or engender violent, immoral, manipulative, and exploitative conduct—it also frequently inspires noble conduct in its devotees.

For the atheist, the view that such conduct is impelled by the fallacy of some future benefit of life after death, the commonly paraphrased Marxist statement "religion is the opiate of the masses" takes on more context. The belief that our life serves as some test for our future life's hopes and aspirations beyond our mortality is often thought to simply serve as a panacea for the futility of life's numerous frustrations for religious cynics. Whether this view is accurate, in large part, depends on any given individual's faith or lack thereof, for no one can truly prove the existence of God.

The impetus for pondering how we should treat others and our environment has powerful consequences. If the greater measure of happiness is the fruitage of the love we feel for ourselves, all beings, and Earth, our efforts to improve our benevolence toward others will ultimately result in improving our own well-being and satisfaction in life. As a society, we don't typically, or consciously, think of success in this way, but we should, and we must in order to find more contentment in life.

Since no other pleasure or pursuit is truly satisfying without love and friendship, should not a considerable investment of our energies and efforts be focused on improving our friendships and relationships? While wealth, beauty, professional accomplishments, fitness, and so on combine to contribute to our satisfaction and well-being, they are only the decorations.

The real foundation for happiness and well-being rests in cultivating the qualities that contribute to harmony and love of ourselves and our families, communities, and environment. Since all life is interconnected, so, too, are all our affections, for in so expanding our hearts, our love grows, for love and hate are the only things that increase the more they are given. We further consider these thoughts on the interconnectedness of life in the next chapter.

Chapter Three
The Interconnectedness of All Life

Spirit. Mind. Body. Family. Ecosystem.

All things are connected, if one pillar is weak, all the others will be affected. All creation is one. What we do to one, we do to the entire web of life.

—Chief Seattle

Since you get more joy out of giving joy to others, you should put a good deal of thought into the happiness that you are able to give.

—Eleanor Roosevelt

I'm a Black dude, and nobody hates being a stereotype more than me, but I really hate the cold. When I moved to New Mexico near a dozen years ago now, I was no stranger to the Southwest. I'd grown miserable in the winters of the bitterly cold Northeast and was determined to move to a warmer climate. I considered Seattle, Portland, Santa Fe, Tucson, Sedona, and Miami. Seattle and Portland weren't much warmer during the winters I'd grown to hate, but at least their winters were milder. I'd always loved their chill artistic vibe, and the outdoorsman culture of the Pacific Northwest and its pristine beauty, so I kept those cities on the list. I wanted somewhere where the cost of living was lower, too, so I could start to make some headway on the student loans I'd taken out in medical school to cover the cost of living expenses, and that eventually narrowed my search.

When I was twenty-five, I took a vacation at the end of my medical internship year to make a Southwest road trip. I've always loved long drives and photographing landscapes, and road trips allow me to get out of my head a little and enjoy America's diverse scenery. Reconnecting with art and nature and savoring some quietness always helps me to work out whatever's trolling my thoughts and peace at the time. And though I don't consider myself a great landscape photographer, occasionally I'll get lucky and capture something of quality if the scenery is picturesque.

I was single at the time, and I did a road trip solo around Arizona and New Mexico. I'd wanted to see the Grand Canyon for years, so I was excited to make the most out of a few days away from my grueling 120-hour-a-week work schedule. That trip forever changed my life, and I've made annual road trips to Sedona ever since.

Sedona is an immaculately beautiful place. If you've never been, it's worth the effort, for pictures simply don't do it justice. Sedona is more beautiful than the Grand Canyon in my opinion, and much more accessible. It's easier to camp, hike, kayak, and mountain bike. Permits are usually not necessary for the locations I enjoy visiting. The more abundant food and the housing accommodations also make Sedona a much more practical locale to visit.

I started seeing more and more of the Southwest each time I'd visit, usually starting in Sedona and then driving all over the Southwest. I eventually settled on New Mexico when the time came for me to choose a new home. I fell in love with the Land of Enchantment's beauty, its art, and its culture, including the rich and tragic history of Native Americans here. I've felt an intuitive bond with

Indigenous people since my early college days for a reason I still poorly understand. Seeing their way of life, their love for the natural world, the harmony of their communities, and the beauty of their awareness of the unseen entities around them made me see life and success in a completely different, healthier, and more balanced way.

My years photographing the poor in Atlanta, Philadelphia, Camden, Cuba, and other places around the United States and Caribbean produced the same results studying and visiting Indigenous communities in New Mexico did. I grew firmer in my understanding that happiness and contentment are not the exclusive possessions of the rich. In many ways, it became abundantly clear that the modern world could learn much from the cultures and philosophies of Indigenous peoples worldwide. Indigenous people's principles contribute to true contentment and harmony within their communities and the ecosystems of which they, too, are a part.

Religion and philosophy have always provided the material most rewarding for me to study aside from medicine. I can hardly remember the last time I've read fiction other than the occasional Dan Brown, George R. R. Martin, or Lee Child novel, whose work I love. There's a rich, intense euphoria, an intellectual ecstasy, that comes from seeing the world in a new, more insightful, and personally rewarding way. Society doesn't always think of religion as an avenue for intellectual study or as material rich for philosophical consideration, but we should. Religion as a philosophy and intellectual field of study is an endeavor as practical as any other. Nothing is more important than understanding human relations and interaction, nothing more practical or useful, and as religion and philosophy focus heavily on such, these subjects are as practical as any other.

As a young man, I wanted to get the most out of my life, to do something that really made a positive difference in the world. I earnestly wanted to achieve something important, something of lasting significance for humanity. I knew my life had a purpose, even if I wasn't completely sure what it was, and medicine had at times tried but failed to pull me away from the creative interests I find most fulfilling.

As I grew, matured, and had a family to raise, practical decisions about how to parent my children became the daily concerns most prevalent in the forefront of my thoughts, more important than any achievement, taking any photograph, or any book I needed to complete writing. Above all, thoughts on the kind of father and husband I really wanted to be, and what I want my children to learn from both my word and example, became my most important priorities. In many ways, it's all too easy to get lost in the habit of simply surviving, that is, simply being a cog in the machinery of society, living to work, and not enjoying or savoring the short time we have. It's a battle I've repeatedly had to fight as the monotony of obtaining the necessities of life, and the challenges of practicing medicine, sometimes seemed to smother my other more fulfilling creative pursuits and familial obligations.

The Bible is a much-maligned book by scientists and intellectuals; yet it is a work greatly esteemed among humankind still—and for good reason. It is the best-selling book of all time, and the beauty of Psalms, Jesus's Sermon on the Mount, the wisdom of Solomon's words in Proverbs, and the book of Ecclesiastes make these and other Bible books masterpieces of literature and treatises of ethical behavior, aside from just their perceived religious value. The Bible also presents antiquated thoughts on many topics often seen as profoundly misogynistic and moral views that present sexually and gender-identity repressive norms. These thoughts have deeply affected humanity's social development and perspective on a host of issues. If for no other reasons than those, the Bible is still a book worth examining, if not simply to comprehend how its popularity and Judeo-Christian ideology have shaped much of Western civilization's legal, moral, and ethical perspectives.

Considering world religions and philosophies always took me away from the rule-based world of the scientifically and medically rational to considering the more important matters of character and ethics, things that we all can feel pressure to compromise in the pursuit of more practical matters in life, or more often simply fail to give any attention to at all. More immediate interests and needs that seem more practical often make it difficult for a person to consider their spiritual needs and concerns. Reflecting on the happiest times of my life, it was not just about what I was doing, but as importantly, if not more so, who I was sharing those moments with. To be surrounded by quality friends, doing work I enjoy, and enjoying the fruitage of my labor are truly the best ways I've found to spend my life. With that as a measure, the lives of countless Indigenous cultures reflect the wisdom of Solomon's words.

Indigenous cultures are often viewed despairingly and considered inferior, as people focus primarily on the surface of Indigenous lifestyles. The perceived material poverty, the increased physical demand of obtaining the necessities of life without certain technological advancements, and the unique physical differences caused by atypical standards of makeup or practices such as intentional scarring often move people to judge these cultures as ignorant and unintelligent, presumptuously and in gross error. But such an assessment is far from reality.

People today often pride themselves on living a lifestyle that leaves a smaller carbon footprint. They eat vegan, live in tiny homes, drive electric vehicles, recycle, and make other changes in hopes of living more in harmony with the natural world. In so doing, they are only imitating what countless Indigenous peoples around the globe have done for millennia. In many ways, the world is starving for the wisdom these so-called primitive cultures rooted their communities' foundation upon, communities that have endured longer than any industrialized nation ever has, and without destroying their ecosystems. Consider the concept of interconnectivity.

Spirit. Mind. Body. Family. Ecosystem.

No man is an island

"No man is an island" wrote John Donne in a poem in the seventeenth century. His words underscore a vital truth about happiness: There is no possibility for lasting happiness without love and connection with others. True happiness is the fruitage of love felt and expressed for others. It's born from connectivity, a feeling of closeness and concern for others and the natural world, experiencing mutual respect and solidarity within one's inner circle and community, and purposefully contributing to the well-being of others and oneself. A heart of love impels one to give of oneself, sharing friendships and love, and finding purpose in service to others and humanity. For men, it is neither inherently effeminate nor weak, as some may suppose, but an elevated way of living that leads to genuine happiness. It does not exclude the acquisition of wealth nor obsesses or sacrifices relationships to obtain such. Genuine happiness is rooted in an internal condition of the heart, the love we feel and express for others and ourselves. It is not dictated wholly by life's circumstances nor the losses and grief invariably a part of living.

Connectivity with Self and Family

Finding purpose in life and understanding oneself is an endeavor that's neither a linear process nor a brief one. As we mature and gain experience, we often find ourselves evolving, enjoying and appreciating new things. For some, the toxicity of a harsh or abusive family upbringing, the fallacy of pursuing various illusions of success, and other challenging or tragic circumstances can set a person on a course in stark contrast with the values and attitudes that lead to genuine happiness. But all is not lost. Regardless of one's start or failures in life, connectivity can be gained and growth can occur in a relatively brief time for any determined individual.

No matter where we are in life, inside us will always be an internal need for belonging, of having and feeling purpose in life. "No fucks given" is an increasingly popular mantra, and even books have been written advocating this perspective. There is a growing popular push not to care about what anyone thinks or feels about anything. But that is an attitude that invariably leads to disconnection and loss of solidarity. While not losing ourselves to the expectations of others, or squeezing our points of view toward servility or our personalities into boxes to meet others' impositions, the need to feel connected with others and serve a greater purpose is a part of our inner being that should never die as long as we live. It is a part of what makes us human; it is fulfillment of this desire and a need for solidarity and purpose that elevates us from the mundane reality of just existing. In caring for others, our ability to manifest love becomes fuller and, ultimately, so does our happiness.

What makes you feel connected? As humans, we tend to focus our energy on the things we enjoy doing. Our pleasure in certain activities is often a clue where our pursuits should guide us in life as a career. As long as the pursuits we enjoy are not self-destructive or destructive of community or family, the happiness we feel while engaging in the activities we enjoy is a beautiful gift. What type of activities help you to feel peaceful? Relaxed?

Happy? Alive? What is your gift? What if you think you don't have any gifts?

When I was young, I wanted nothing more than to be like my dad. Forty years older than me, and with a potbelly, my dad had neither abundant artistic nor significant musical or any other creative talents. Though trained as a mathematician, he worked as a computer programmer. He spent his mornings reading the *Washington Post* comics section first, then did its crossword puzzle and sudoku. He would check my homework, talk to me at the dinner table, and support my athletic interests, which were my primary recreational interests as a child. Nothing incredibly inspiring was happening on the surface. But I loved him, and I grew to appreciate and respect him because of his internal gifts and his genuine care and concern for me.

My father is a brilliant man; he is also the most gentle and humble man I've ever known. When I was a young child, it wasn't his physical or artistic prowess that affected me, it was his reliability and kindness. Despite his brilliance and professional accomplishment, I've never heard my father raise his voice, be condescending, or treat anyone with less than kindness and respect. A person might incorrectly assume him a pussy, watching him handle some unkind treatment from a stranger with humility and restraint. But he is no weak man. An ex-soldier, six foot two, with piercing eyes, his physical presence alone was often enough to keep would-be swindlers at bay. One look was often all that was needed to set some offending stranger on his heels.

Those were some of the experiences that affected me the most as a child, and I remember making a conscious choice to pursue restraint as my father did. It was not because I had any real appreciation of what humility really means, or how his commitment to religious devotion inspired such, only that I appreciated his kindness. It was an internal gift, inspired from his religious devotion, and it no doubt contributed greatly to his long professional career without complaint and no question to his persistently calm and pleasant disposition.

My mother is the more outwardly talented of my parents, but she, too, has beautiful internal gifts. Gifted with exquisite penmanship and a powerful singing voice, she could easily, with the proper support and instruction, have been an opera singer. While I liked her voice as a child, it was her generosity and emotional intelligence I appreciated most about her.

I recall my parents' effect on me to illustrate an important point: Many individuals feel like they have no natural gifts or talents, and perhaps they may not. But the absence of unique skills doesn't preclude one from having a life full of healthy relationships—nor does the possession of extraordinary natural ability guarantee friendships and love. In fact, it is not uncommon for inherent abilities to incline individuals toward internal qualities like arrogance and inflated feelings of self-importance, qualities that often inhibit the cultivation of close relationships.

Few things foster friendships and bonds of affection like sharing and doing what we love with others who share a common interest in similar activities. If we have a family, our generosity most easily manifests at home if our hearts are inclined appropriately toward benevolence, and our love will innately find expression first toward our closest associates. Our mate and children naturally are the immediate beneficiaries of our efforts to improve connectivity with them. Even if our children and our partner do not share our same passions or interests, would they not still appreciate our sincere efforts to bond and draw closer with them?

Understanding the Pillars of Connectivity

If you asked my wife to make a list of the things she'd like to change about me, I can guarantee you somewhere near the very top of that list, if not at the very top, would be the length of time I spend on electronic devices. I love my phone. It's the perfect companion for my ADHD—on any given day, I can guarantee that blitz chess, Zooba, or Apex Legends will be on my unofficial to-do list for at least an hour. I love to write, but I usually write either late at night or early in the morning when the children are asleep. Years of working alternating night and day ER shifts have destroyed any regular sleep patterns I could ever hope to have, and I've learned to get my writing down when I can. The same goes for sleeping. You have to punch the keys when the inspiration strikes. The reality of balancing life, work, recreation, fitness, and helping my wife, at least to a degree both she and I can reasonably find respectable, is not always an easy task.

That sophisticated juggling act, I'm sure, is even more difficult for my wife and countless other mothers.

Like many mothers, my wife is definitely the go-to parent for any number of nonemergent emergencies our children never fail to present throughout the course of any given day. Finding happiness is not so much a goal as merely finding peace of mind against the litany of to-do activities that inundate us. We can't find the strength to pursue lofty goals of happiness and enlightenment unless we can manage the necessary tasks of our day-to-day lives. Taking a moment to ensure our emotional well-being is cared for often makes the difference between responding with kindness and patience versus anger and sharpness both at home and in the workplace.

People use various techniques to cope with stress. Finding what works for us individually is more important than putting on something that works for someone else but may not necessarily be the best fit for us. I don't presume to be an expert on what works best for others, but there are things I know that work for me, along with other data-proven strategies worth considering that can help us to achieve optimal mental health and fitness.

Large segments of nearly every medical and mental health specialty espouse the virtues of physical fitness, meditation, prayer, walking, hiking, gardening, and other hobbies and activities. But people are diverse and have different interests. In New Mexico, for example, in the mostly rural community where I live, a large segment of my community enjoys hunting, rodeo, fishing, motocross, and firearms. I experienced culture shock when I first moved here from New Jersey. In Jersey, my friends and I usually went to the gym, played soccer, or enjoyed a game of Risk or FIFA. While hiking wasn't uncommon, camping was certainly an extravagance for us, and rodeo was surely on my I'll-definitely-never-do-that-shit list. But life has a way of throwing us curveballs so that, if we dare to have the courage of an open mind, we might find ourselves pleasantly discovering something we'd never imagine we'd consider.

In more rural parts of New Mexico, away from cosmopolitan Santa Fe, rodeo is king. There are adult and youth leagues, and children, sometimes as young as three, often participate. If you've never seen a poised three-year-old ride a bronco or stallion, it's an amazing sight not easily put into words. The first time I saw the families at these youth rodeo events, I felt compelled to do a photo-essay about them. Seeing small kids roping, barrel racing, and completing other events, and seeing their families and volunteers supporting them as rodeo clowns and other supporting staff, is a story I just had to tell. Every community worldwide has its special events and beautiful things—and New Mexico is no different.

Ruidoso is breathtakingly stunning and has a large Mescalero Apache reservation and casino. Santa Fe has its art, culture, and fine dining. Silver City has the grand Gila National Forest and Cliff Dwellings, and Chaco Canyon punctuates the state's central western area. An assortment of large and picturesque lakes and the Bosque Del Apache Wildlife Preserve host hundreds of thousands of migratory birds each year. Embracing these local gems helps connect me to my community and its history and fosters a connection to the land sustaining me. I have to maintain at least a reasonable degree of fitness to play with my children and enjoy the outdoor recreation I enjoy, and that helps me at least achieve a measure of fitness. Many cultures embrace the concept of spiritual harmony, mind, body, family, and community connection. Each individual may have unique ways of achieving this balance, but a deficiency in any of these areas is sure to result in various unpleasant consequences.

Spiritual harmony

The central principle present throughout many Indigenous cultures is the concept of universal harmony, that is, living in peace with all beings, both animate and inanimate, with a conscious commitment to contributing to the well-being of everything surrounding us. The impetus is on maintaining balance with all things—not just interpersonal peace and harmony with other humans. Achievement of such balance requires constant effort and awareness that all life is in a continual state of flux. Our attention and focus must be persistent and undivided to maintain the habits of principled stewardship required to attend to the continually changing needs of our family and the people and ecosystems surrounding us. This balance must start first within us individually, with an intentional effort to understand ourselves, our place within this balance, and our personal needs, weaknesses, and gifts. Cultivating a commitment to achieving our individual best selves means providing sustenance for our spirituality, mind, body, family, and community in a way that shows respect for nature's intricate cycles and the interconnectedness of all life. This is our sacred commitment to our ancestors, without whom we would not exist, as well as to our descendants, who will inherit the land and circumstances we leave them.

Mind

To achieve our optimal personal development and maintain the focus needed to persist willingly in the stewardship of all that surrounds us, our thoughts, principles, and actions must all be in harmony. Attentively caring for the needs of ourselves and others requires disciplined self-sacrifice and unwavering focus that can only be fueled by a mind that is strong and focused. The power derived from this personal philosophy helps us to have the fluidity and fortitude needed to manage the inevitable changes inherent to life and the unwavering commitment to unyielding personal morals despite the inconvenience and sacrifice such commitment might mean for us personally. An awareness of the greater good, that which is most important, steadies our aspirations and choices—not just with a focus solely on what is in our immediate individual interests but what is in the long-term interests of our family, community, and ecosystem.

Body

The ability to achieve the mental fortitude required to be conscientious stewards of our ecosystems mandates an internal strength that can be augmented by the self-discipline and physical endurance physical activity engenders. This mind–body connection principle is present throughout many cultures. Shaolin monks cultivate one of the most respected martial arts traditions in history and are known for their unique dexterity, strength, and agility. Much of their commitment to martial arts is motivated by an awareness that physical discipline augments their spirituality. For example, being physically fit was observed to help monks achieve more effective meditation and avoid falling asleep during this time. Many Native American and Indigenous cultures feature coming-of-age rituals that include various tests of physical strength and endurance, prowess as hunters, and other survivorship tests. Physical strength is integral to survival in many Indigenous communities, where the threat from other humans, more lethal forms of wildlife, and other environmental factors is ever present.

Family

Family is a sacred bond—and it's not just those with whom we share our blood or genetics, but those, too, with whom we share intimate bonds of platonic affection. One of the characteristics that elevates humanity from other forms of animate life on Earth is that our bonds of affection extend further than to just those with whom we share a genetic bond. "There is a friend sticking closer than a brother," states one proverb, and that is a truth that extends throughout every part of the globe, despite cultural and religious variance. For many Indigenous peoples, there is an ever-present awareness of the sacredness of the familial bond. The indelible link and debt we owe to our ancestors as well as to our descendants is ever present. This principle engenders a continually conscious awareness of our responsibility to care for every aspect of our being mentally, physically, and spiritually and extend that commitment of stewardship to those we are most intimately associated with, namely, our family and inner circle of close associates.

Ecosystem harmony

Our immediate community are the families and individuals with whom we share geographic proximity. That is the most rudimentary form of community organization. Globalization has made the interconnectedness of humanity more intimate; in actuality, all of humanity is our family. Still, the direct sharing of geographic proximity will always define our most immediate community, as humans are animate beings who take physical form. As such, we theoretically share every resource of the community of which we are a part. The water, air, land, wildlife, mineral, and other resources must be protected and shared with the awareness that harmony dictates. Greed as well as wanton wastefulness must be avoided, for the balance between humanity and the natural world is a delicate one. We take what is needed, but not with misanthropic asceticism, for when shared, the bounty the natural world provides far exceeds what is necessary for survival. The foundation for every manner of artistic expression is provided generously in the abundance of the surrounding mineral, animate, and inanimate resources of the natural world.

Deficiencies in the Pillars of Connectivity

Spiritual disharmony

Rampant spiritual disharmony is readily apparent in our world. A lack of appreciation for the interconnection of all life is displayed in the grotesque abuse of the natural world that sustains us. This disharmony is principally the result of the greed of a few, the masters of mankind, who propagate cycles of consumerism that enslave the bulk of humanity through a variety of psychologically indoctrinated values, behaviors, and habits. These cycles of never-ending consumerism engender wanton exploitation of the environment, relentlessly harming the ecosystems upon which humanity's survival depends. Reinforced through the lack of connection with the natural world, humanity's reliance upon a small segment of the populace for its food production has resulted in generations of humanity psychologically disconnected from the actuality that these ecosystems sustain our existence. We go to the store, buy our bread and food, turn on the tap or grab a bottle of water, and completely forget that everything that sustains us originates not in a grocery store but directly from the only source capable of providing for our physical nourishment: Earth.

    This proliferation of the cycles of environmental disconnection is one of the principal consequences that the development and sustenance of cities wrought upon humankind. For humanity to break the cycle of spiritual disconnection globally is a task I'm not convinced is possible except through generational change or global catastrophe. Adults typically have entrenched habits that contribute to ongoing environmental decline and disconnection to the natural world, and the influence that the masters of mankind hold over society is so encompassing, without a cataclysmic shift in the way we educate our children or a complete change in the global order of society by unforeseen circumstances, the changes required to overcome generationally entrenched consumeristic habits is unlikely to occur organically.

Without a significant proportion of humanity contributing to the production and maintenance of their own food and water supply, the bond humanity feels with Earth will continue to be insufficient to make the changes needed to adopt lifestyles that support a sustainable relationship with the natural world. The only path that will ensure our species' survival and preserve the delicate ecosystems of which we are a part is for humanity to feel a meaningful connection with Earth. Such is needed to implant and sustain habits that show proper respect for the delicate balance of the life cycles on our planet and within its ecosystems. When every individual contributes in part to the cultivation of their own community's food and water supply, only then will society understand and respect the obligation we owe Earth and our descendants as stewards of the natural world.

Such an enormous shift in humanity's manner of life is impossible under the fractionated governments we now know. The petty division and sectarianism that dominate human civilization represent a gross lack of moral development and enlightenment in humankind as a species. There is an ever-present need to establish a global government, but I am not of the mind-set of some ascetics who call for a complete abandonment of our technological advancements.

The development of communities that incorporate innovative architectural design and planning, allowing for water conservation and significant land allotments for community crop cultivation, is not a mere fantastical thinking dream. Solving a fundamental problem doesn't mean we have to approach it in unsophisticated, aesthetically unappealing, and uninteresting ways. Some of the most beautiful things in the natural world also serve incredibly practical functions. There's no reason sustainable architecture and city planning cannot embrace the same beauty and functionality readily present throughout the natural world, addressing our needs in elegant, aesthetically pleasing, and environmentally conscious ways.

Mind disharmony

In no small degree, the philosophy that dominates our thinking and heart is a combination of our education and the indoctrination we receive through examples as children. The values instilled through educational development, in the Western world especially, focus little on the cultivation of patterns of thinking and habits that are in harmony with sustainable living. Instead, emphasis on individual success fostered by competition is the primary way most humans perceive success. Principles of Indigenous peoples are often discarded in Western civilization out of ignorance of these principles' value.

While Indigenous cultures often never developed the technology standards of the Western world, they were, and are not, ignorant people. Their manner of life reflects recognition of that which is most important, namely the ethical stewardship of Earth and the protection of one's ecological community and family. Typically, there is often no concept of success in many Indigenous communities if the community as a whole is not prosperous. The individual is placed second, the community first. These principles must be instilled in humanity globally, without disregarding advances of technology—rather, by incorporating, developing, and implementing technology in a way that demonstrates a responsible and sustainable way of life for humanity and the ecosystems of which we are a part.

Bodily disharmony

Without health, every joy is diminished and every success less sweet. Our physical body justifiably should be cherished, and every effort to attain a measure of fitness that enhances the possibility of longevity should be pursued. Such is not especially easy for some, as the demands of family life, work, and other responsibilities often can make pursuing fitness a challenge. Overcoming habits of inactivity, smoking, heavy alcohol use, poor nutrition, and other unhealthy habits requires internal commitment and diligence that are not easily obtained without disciplined persistence.

    Individuals must fight to find the willpower to make such changes, and there is no shortage of motivations that empower people to do so. For some, fitness is rooted in vanity, the desire to be as attractive as possible for as long as possible. Vanity is a powerful motivator—and not without good reason, for many opportunities are extended to attractive individuals that others less comely cannot gain, both romantically and professionally. Many are commonly motivated by a desire to remain appealing to the individuals they've made a commitment to romantically and understand that, in a committed relationship, an effort to maintain their appearance is one of the ways they show devotion to their partner.

For some, a commitment to their children is the primary motivating factor. That was the primary factor that helped me quit smoking. I would look at my youngest daughter and think about not seeing her wedding day or meeting my potential grandchildren. Everyone is different, however. What works for some may not be enough for others. Some find the strength through prayer and meditation to make the changes they know deep inside they need to make to improve their health.

Not the least among the motivating factors for some is simply a desire to achieve the life they know they really want. When our thoughts repeatedly pace us to make certain positive changes in our lifestyle and relationships, it's an indication of the direction toward which we should move. We must find a way to heed our intuition's prodding and common sense's admonition. Happiness and health are often the rewards. Being physically fit is a habit that yields beautiful fruitage. Feeling better, having more energy, being healthier, being more attractive, having increased flexibility, and enjoying the pleasures of being outdoors savoring nature, be it through walks, skiing, hiking, surfing, walking, or any number of activities, are worth the sacrifice and effort it takes to achieve such. Instilling the habit of physical activity in our children is also a goal that is worth pursuing, for such a routine can help contribute to a child's physical and emotional well-being for a lifetime.

Whatever motivates us as individuals, we are alive on this Earth for a very brief time. Being physically attractive, having better health, better flexibility, and the ability to enjoy other benefits from fitness enhance our ability to be mindful stewards of Earth, which undoubtedly makes caring for our bodies a worthy endeavor. I don't have all the answers to what motivates individuals. Still, I know that people find the strength and time to pursue all manner of activities that are less beneficial than taking care of themselves physically. If people can find the motivation to pursue things that offer them little to no lasting benefit, they can find the motivation to pursue activities that will make a significant difference in the quality and enjoyment of their life.

Family disharmony

Family is a sacred bond, and an indelible one. For some cultures, family extends well beyond the nuclear family. Families can be an amalgam of friends who are such a part of our lives that some can even feel closer in bonds of affection than family. It can be easy to take those closest to us for granted at times. The burdens and challenges of earning a living and pursuing various goals, interests, and hobbies can require significant commitments of our time. Sometimes, we can fail to appreciate that the happiest moments in our lives are often those simple joys we share with others. What might be gained by sacrificing our time spent with family and friends might result in the weakening of those very relationships that bring us our greatest measure of happiness.

What we feed and give light and affection to grows. It matters not if such are goals, anger, fitness, love, relationships, plants, or anything else. Sometimes it takes the loss of a relationship, or a mistake that changes the course of our lives, to truly appreciate our health, job, some friend, or lover or something else that we have undervalued. When we look at cultures often considered inferior, that is, cultures that place value on spending time with extended family and friends, we often judge them by what they lack and not by what they have. The lack of technological advances speaks nothing about the possession of close relationships and friendships, and relationships are the most valuable assets in life. Often, the sweetest moments in life are woven with the mundane, for example, the walks, the sharing of meals, and the playing of games.

In these simple moments, the bonds of friendship are often sown and solidified, and the path toward love and community is built on those inconsequential activities, which show us more profound truths about the people we grow to care for. Sometimes in the pursuit of goals and other interests, we necessarily forfeit some things of value. Still, there is a difference between conscientiously choosing to sacrifice certain things versus losing things we value because we neglect them or fail to appreciate their importance.

Leaving New Jersey for New Mexico was not an easy decision. I'd made many close friends in Jersey after spending eleven years there, some of whom were closer than even my family in some respects. When I left, I knew that those relationships would never be the same, but I also felt that the move was necessary for my life's journey. It was something I felt had do to achieve something of significance for society and for myself. For years, I'd felt like I needed to become a writer. I would work on projects and not finish them, or fail to devote the time to writing I needed to because I'd rather just hang out with my friends. I loved my friends, but, deep down, I knew spending my evenings after work and my weekends playing pool, Risk, Monopoly, FIFA, or soccer, and having empty romantic relationships, wasn't the course I was meant to live. I felt an intuition inside pushing me in a different direction. I knew for me to achieve something of significance, something I felt I was meant to achieve, I had to sacrifice spending so much time in recreation.

Sometimes people have drives and goals that others cannot fully understand. The need to be an artist, actor, comedian, writer, musician, build a business, and so on is something many people cannot relate to because the gripping drive of being a creative or an entrepreneur doesn't reside in every person. Even well-intentioned friends and family can misunderstand those who feel the drive for greatness—not only in the arts but in other fields like business or science and technology.

Depending on the part of the world in which one lives, there can very defined goals impelled by stereotyped gender norms. Little girls particularly can have interests in stark contrast with gender-defined stereotypical hobbies and pursuits society often indoctrinates. It is an ignorant assessment to consider the intuitive drive some individuals feel about pursuing certain interests as a sign of irrationality or delusion. Many of the most successful entrepreneurs and artists of all time saw their path and destiny long before others could see their talent or understood what in time they would become.

There is an intricate balance between making a living and pursuing goals and balancing family, friendships, and romantic relationships. The pursuit of purposes related to providing materially for ourselves and family can sometimes distance us from the very people we love the most. But pursuit of such does not necessitate the abandonment of these relationships. However, it often means we must find new ways of cultivating closeness with those we love, especially if our path takes us away from them either figuratively, literally, or both.

Technology has improved the means for maintaining relationships with those we love in spectacular ways. FaceTime, Skype, WhatsApp, TikTok, Snapchat, and other social media apps make sharing our time and lives with others more convenient than ever. Still, nothing will ever entirely replace the beauty and simplicity of sharing a meal or walk with a close friend or friends, family, or a lover. More important than the activity is the effort we demonstrate to ensure the bonds of affection we feel for those we love are not broken. Sincere effort is a noticeable thing, often just demonstrating that we sincerely care is as important as anything else to remind those for whom we feel affection they still live in our hearts and minds.

When I left for New Mexico, my friends called, texted, and video-chatted with me, and I did the same. They visited me, and I visited them. Some married, others had children, and I was right there and held their children in my arms and even changed the occasional diaper. Eventually, I married and had children of my own, and my closest friends were right there by my side.

Every family is unique, as is every child. Only through attentiveness and routine communication can we have an idea of what the ever-changing needs of those we love are. We owe an obligation to be our best selves to ourselves primarily, but also to those who depend on us for support financially, emotionally, spiritually, and otherwise. To make every effort to guide, support, and provide for their needs, as long they are reasonable and healthy requests, is among our most sacred of responsibilities. In part, our obligations are diverse, depending upon the needs and circumstances of those we love, but still with an ever-present cognizance of the need to be an example in both word and deed of the qualities we want most to see manifested in others.

Ecosystem disharmony

As climate change reveals the patent disharmony between humankind and the natural world, the innumerable violent conflicts and bloodshed within human civilization reveals the flagrant interpersonal disharmony humanity experiences as a species. No part of the world is without violent conflict in one form or another. In some places, the violence occurs on a larger scale, manifested by civil wars and genocides. In other locales, violence manifests primarily through crime. While conflict is unavoidable, violent conflict almost always is. The proliferation of large-scale violence in the face of society's strong opposition to war displays the power the masters of mankind have to shape the course of humanity.

Aside from war and the gratuitous violence some experience through random acts of malevolence, a pervasive air of division and disharmony permeates society. We can feel it in the way people interact with each other. People immediately assume individuals have ulterior motives when they are kind, and such skepticism itself is a sign of a grossly dysfunctional society—and not without reason. The proliferation of every manner of deviant behavior is present throughout society, and it doesn't exclude people in positions of trust, for example, teachers, priests, doctors, police officers, and so on. It is not surprising that serial killers, rapists, con artists, thieves, drug addicts, and other individuals who pose a danger to public safety have produced a society in which its members routinely greet each other with an undercurrent of mistrust. The concomitant existence of racial prejudice and social class distinctions augments this reality, and the illusion of superiority is easily recognizable in the degree of fracture and mistrust that permeates society.

But how do we heal as a society? In the upcoming chapters, "The Failure of Education" and "The Inevitability of Global World Government," I speak in detail about strategies that can help humanity heal various divisions and challenges we face together. One thought I'll mention here is the concept of intracommunity cultural exchange.

The Value of In-Person Human Interaction

When I first saw the Obama Presidential Library architectural plans, it immediately reminded me of an ancient world cultural and economic center like ancient Mali's Timbuktu Library and the Library of Alexandria in Egypt. There are countless examples of similar cultural, educational, and financial centers throughout the ancient world. Still, when history is taught in America, such places' histories are often excluded, especially if they were found in non-European cultures.

Consider ancient Timbuktu, which was founded around 1100 CE, and, in time, because of the city's strategic location twenty kilometers (12.4 miles) north of the Niger River in Mali, it became a sprawling hub for the sub-Saharan gold and salt trade. After making a pilgrimage to Mecca in 1324 CE, Mali emperor Mansa Mūsā built the Great Mosque, Djinguereber, and a royal residence. Later, Musa commissioned Sankoré University by a Granadan architect, and, by 1450 CE, Timbuktu's population increased to about 100,000, with 25,000 scholars, many of whom were products of Meccan or Egyptian tutelage.

Timbuktu and Egypt's renowned Library of Alexandria were economic and educational cynosures, attracting commerce and scholars from diverse locales around the world, and many Greek scholars journeyed to Egypt for educational and scholastic reasons. President

Obama's Presidential Library's architectural design pays homage to those ancient centers of commerce and scholarship, and their example as centers of enlightenment and prosperity provides a path forward for modern civilization that is worth imitating.

The division in America is one of educational and economic differences as much as it is one of racial divide. An additional factor intricately intertwined with these is the inherent differences and divisions between rural and urban centers. The polarization within news media outlets and social networking algorithms often reinforces the biases ingrained in society. The advent of internet-based entertainment and social media platforms has in no way expanded the diversity of thought and experience needed to incorporate variety within the American educational experience but, instead, through algorithms specifically designed to align with users' interests, has contributed to reinforcing established prejudices and political views. The need for an in-person interchange of commerce, educational, and recreational experiences is more vital than ever to produce the diversity in cultural exchange that inevitably leads to less bigotry and prejudice, as only direct interpersonal exchanges can provide.

We learn the true ignorance of prejudice only through exposure and understanding of peoples, cultures, and ideas that look and think differently than we do yet who still share our same basic joys, needs, and aspirations. Much bigotry, even in deeply ingrained individuals, is often overcome through sport, the sharing of meals, and the exchange of commerce with each other. Malcolm X, a man who once both owned and eloquently espoused deeply entrenched and grotesquely bigoted views, wrote in a letter of his transformation and recognition of the profound ignorance of his prejudice and racial hatred after his hajj to Mecca. His direct, in-person experience completely altered his previously bigoted views. In the letter regarding that experience, he wrote,

> There were tens of thousands of pilgrims, from all over the world. They were of all colors, from blue-eyed blondes to black-skinned Africans. But we were all participating in the same ritual, displaying a spirit of unity and brotherhood that my experiences in America had led me to believe never could exist between the white and the non-white.
>
> You may be shocked by these words coming from me. But on this pilgrimage, what I have seen, and experienced, has forced me to rearrange much of my thought-patterns previously held, and to toss aside some of my previous conclusions. This was not too difficult for me. Despite my firm convictions, I have been always a man who tries to face facts, and to accept the reality of life as new experience and new knowledge unfolds it. I have always kept an open mind, which is necessary to the flexibility that must go hand in hand with every form of intelligent search for truth.
>
> During the past eleven days here in the Muslim world, I have eaten from the same plate, drunk from the same glass and slept in the same bed (or on the same rug)-while praying to the same God with fellow Muslims, whose eyes were the bluest of the blue, whose hair was the blondest of blond, and whose skin was the whitest of white. And in the words and in the actions and in the deeds of the 'white' Muslims, I felt the same sincerity that I felt among the black African Muslims of Nigeria, Sudan and Ghana.

We are *truly* all the same-brothers.

All praise is due to Allah, the Lord of the worlds.

The proliferation of internet commerce has aided in society's globalized segregation by reducing the degree and prevalence of direct in-person interchange. A concerted investment in commercial endeavors that foster direct interpersonal interchange is essential, for in every interaction of the personal—be it commercial, education, recreational, spiritual, or otherwise—the seeds for understanding can be sown and cultivated. This is aside from the economic benefit of protecting brick-and-mortar small business interests by cultivating in-person trade of goods.

President Obama's Presidential Library designs birthed within me a realization that one of the paths forward for society is revisiting what these combined commercial, educational, and cultural centers in the ancient world represented. They were the birthplaces for intercontinental learning and interaction, and humanity must make every effort to encourage diverse in-person exchange if it is to break the cycle of deepening globalized segregation that has grown in recent decades, despite society's advances in technology. Community centers that foster in-person interchange between diverse communities, including rural, suburban, and urban communities, offer a potential path toward societal reconciliation.

Centers offering state-of-the-art recreational facilities and unique direct commerce opportunities for merchants, artists, agricultural, and other fields would bring together truly diverse populaces. Job training, spiritual and technical educational activities, health care facilities, and other industries could be integrated alongside recreational and entertainment facilities, enhancing the appeal and usefulness of such facilities, garnering the appropriate support and attraction required to produce significant-size audiences that would actively participate. Subsidies provided by the government and funding from private industry to support such centers could offset citizens' cost to enjoy such facilities.

The incorporation of various volunteer-based organizations, like physician exchange organizations; in-residence artists' programs; English and other language learning programs; and job training programs would attract individuals from various educational and economic statuses and ethnic diversities. The appeal of state-of-the-art physical training and sports facilities and the formation of athletic leagues with appropriate public venues could feature entertainment that could also support professional-level athletic events.

The goal is to provide the very best low-cost-to-the-public educational, commercial, athletic, spiritual, and entertainment opportunities, including volunteer opportunities for the untrained, undertrained, and upper echelon of every field. Using state-of-the-art facilities subsidized by the government and private industry, the impetus for those of acclaim to periodically volunteer their knowledge and skill would provide encouragement in benevolence, and all would benefit, for none lights another's path without brightening their own.

While the cost of facilities would undoubtedly be enormous, the return for society would be worth such investment, for what is of more value to society than for its citizenry to be less prejudiced, more understanding, more educated, more prosperous, more satisfied, and more peaceful? Creative programs to attract young physicians, teachers, and experts in all arenas through tax incentives, loan repayment, and other economic incentives such as quality housing would attract the finest talent globally. What other nation would be more well equipped than America to reveal the beauty of its commitment to its founding principles by encouraging diversity and the integration of multicultural, educational, economic, spiritual, entertainment, and recreational centers? The true essence of a community center is a place where technology is embraced but interpersonal interaction is prioritized and the interchange of multicultural multidisciplinary endeavors is epitomized and celebrated. Such centers would become the foundation of a more inclusive and connected society.

Whatever path we choose, a great many things must change if we are to incorporate a way of life that is sustainable with the natural world and between humanity itself. What we have done and are doing in Western civilization is not sustainable. Our world is spiraling toward catastrophic environmental collapse. Without a dramatic alteration in interpersonal relationships, and our relationship with the natural world upon which our survival as a species is dependent, cataclysmic loss of human life and biodiversity will be the inevitable outcome. Some speak of those compelled to fight for ecological conservation as alarmist. But, in truth, the alarm about what human activity has done and is doing to our planet is not sounded enough. The condition of our planet's ecosystems merits constant discussion of the realities of the dangers and challenges humanity faces. Instead, they are often an afterthought, for there still remains considerable pushback about the veracity and accuracy of the accounts regarding our planet's environmental decline from various political agents fueled by the economic interests of the masters of mankind.

As a species, we must hold these masters and their corporations accountable. They control and manipulate society for their own economic interests, despite knowing the lasting harm to humankind and the natural world they perpetuate—and that is inexcusable. The survival of our species depends upon the ability and strength of the upright among humanity to dictate the course of our civilization's destiny so that the will of the people may be directed toward patterns of behavior that are sustainable —and not dictated by the economic greed of a select few.

Chapter Four
Hidden Pearls: Jewels of Indigenous Philosophy

Being Indian is an attitude, a state of mind, a way of being in harmony with all things and all beings. It is allowing the heart to be the distributor of energy on this planet; to allow feelings and sensitivities to determine where energy goes; bringing aliveness up from the Earth and from the Sky, putting it in and giving it out from the heart.

— Brooke Medicine Eagle

Some of you think an Indian is like a wild animal. This is a great mistake.

— Chief Joseph

The land is sacred. These words are at the core of your being. The land is our mother, the rivers our blood. Take our land away and we die. That is, the Indian in us dies.

— Mary Brave Bird, Lakota

Cohiba is now an internationally known cigar brand, but it was born in Cuba. It gained fame as one of the finest cigar brands in the world because of the pristine quality of its tobacco blend and its exceptionally aromatic aroma. It was the preferred cigar of Fidel Castro, and on every box of Cohiba is a picture of a sixteenth-century Indigenous chief named Hatuey. Spaniards arrived on his island of Hispaniola (modern-day Haiti and the Dominican Republic), first Columbus and his men and then an estimated one hundred Spanish settlers, soldiers, and an additional estimated one hundred African slaves. The settlers and soldiers subjugated the Indigenous population there, enslaving them, stealing their gold and land, and raping Indigenous women. The African slaves frequently escaped their Spanish enslavers, finding refuge among Hatuey and his people. Then, the Indigenous people, with the escaped slaves beside them, engaged in open violent rebellion against the Spanish, many of whom they killed. Enraged by the Spanish invaders' brutality, Hatuey left Hispaniola in a canoe with approximately four hundred of his men to warn the Indigenous of Cuba about the threat the Spanish represented. Hatuey helped the Indigenous in Cuba resist the Spanish there for a year but eventually was captured and burned at the stake after being betrayed by one of his men.

Before his death, a Catholic priest asked him if he wanted to accept Jesus so that he could live in heaven. Hatuey refused, declaring that if the Spanish were in heaven, he would rather go to hell. Hatuey's rebellion is considered by some historians as the first act of a freedom fighter in the Americas. He is a fitting symbol for the Cuban-born Cohiba cigar brand, as Cuba is a nation with its own rich history of rebellion and freedom fighters, and Hatuey himself once walked, fought, and died on Cuban soil. His story is not unique, and the world's history of European colonialism, American slavery, the Native American genocide, and Mexican Indigenous genocide all reveal the depths of depravity the human species is capable of.

Around the world, Indigenous people have faced every manner of atrocity and subjugation, principally at the hands of persons of European descent. This fact has not gone unnoticed. For a considerable period in his life,

Malcolm X adopted the belief that "the White man is the devil." His examination of history's repeated atrocities committed by persons of predominantly European descent convinced him of the accuracy of that view when he was young. But his views changed drastically after making his hajj.

Europeans are not alone in history as the perpetrators of grotesque acts of violence and subjugation. There have been atrocities of every manner of evil and perversion perpetrated by every nationality of people on Earth. Because European nations have been the preeminent world powers in the West for the past several hundred years, and writing and documentation have advanced considerably from the time preceding Europe's rise to global dominance, these nations' more modern deeds are more well known. But they are no different from nations who've ruled the world throughout history. Egyptians, Assyrians, Babylonians, Mongolians all had considerably sized empires that ruled over large segments of the world, and they each had their vicious forms of torture and slavery. Race has no bearing or monopoly on viciousness.

Without exception, as every nationality has sown their evil deeds, beautiful things have blossomed from the heart of every people and culture in the form of ideas, philosophies, art, and ingenuity—and Indigenous people are no exception. The world has frequently dismissed and portrayed the Indigenous as ignorant people, with nothing of significance to offer humanity, but this is the furthest from the truth. Countless varieties of Indigenous peoples have lived in harmony with the diverse ecosystems surrounding them for millennia. Our species, bereft of these cultures' universal appreciation and skill in sustainable living, must adopt not necessarily their skills, some of which could undoubtedly be augmented by modern technology, but, more importantly, their patterns of thought. These cultures' and peoples' commitment to being stewards of the environment has gravity for our species' survival as a whole.

Understanding Indigenous philosophies, principles of education, and sustainable living patterns represents a moral educational foundation. These principles form the foundation of conscientious technological education and promises to instill the ethical guidance currently critically lacking in Western educational systems. Such ethical guidance is direly needed to compel the moral use and development of technology and encourage patterns of living that reject unrestricted exploitation of Earth's natural resources. One must not flippantly disregard these cultures' values as has been so ignorantly done in the past.

Indigenous Sustainable Living Principles

There is no such thing as a universal Indigenous philosophy, for there are a great many diverse beliefs among the enormous varieties of Indigenous peoples both living and dead. Still, there are common principles among many of the Indigenous. These principles are not merely worthy of our sincere consideration but represent a nobility of thought that parallels even the richest of Western philosophical traditions more popularly known worldwide. Ideas similar to what we commonly see in Christianity, Buddhism, Judaism, Islam, Hinduism, and other religious traditions are present, as are different, unique perspectives on living that have customarily grossly been under appreciated. Indigenous principles offer a guide to a more harmonious relationship between humanity and the natural world, which no degree of scientific and technological progress without proper ethical and moral restraint will ever achieve. Let's consider some of them.

The seventh-generation principle

In the Americas, the seventh-generation principle is believed to have come from the Iroquois Native Americans, but it was present in many African Indigenous peoples. This principle's foundation is for individuals and communities to think about how the decisions they make will affect seven generations of their children and their communities seven generations into the future (about 140–200 years). The principle is meant to provide a long-term view of important decisions to help people decide whether such a path is the course of wisdom.

The wisdom of this principle is obvious. Imagine if humanity embraced this principle when developing and implementing fossil fuels or fighting against climate change? When considering decisions such as whether to go to war? Is the cost of pursuing an invention worth the cost to humanity and the environment? Are the merits of complex personal choices or goals worth the risk or sacrifice? The profundity of the wisdom of this principle is obvious. Instilling this value in our children has indisputable merits. Should such a focus be ever present in the vast majority of humanity's minds and hearts, how might our world be different?

Walking softly on the earth

"Walking softly on the earth" is an expression advocating the need for an individual to connect with the land upon which they reside as well as an admonition for humility. This process of developing a connection with the land is nuanced and diverse. It involves having a conscious and humble recognition of the people who lived upon the land before you, how they preserved it for you, and how you must, in turn, preserve it for those who will live on after you. It is a call to humility and respect of one's environmental footprint, its ramifications for the ecosystem of which one is a part, and the need to maintain humility in one's approach to living and learning.

The interconnection of all life

Holistic thinking. For many Indigenous peoples, all things are connected—not just all living things but all inanimate objects (see "Interview with a Native American Shaman," a sample from the upcoming book *On Spirits,* at the end of this book). Mountains, rivers, rocks, streams, plants, animals, insects, and humans are all bound through a sacred connection as shared parts of the universe. All things share the same universal energy. Humans are just one piece of this beautiful universal ecosystem that comprises both the inanimate and animate. With the awareness of all things' interconnection and this shared bond of energy, there comes an inherent responsibility for humankind. We are the stewards of everything organic that surrounds us. All beings, and all things, are all essential pieces of the delicate web of life, including unseen entities, which nearly every Indigenous culture recognizes. Our sacred duty rests in the way we honor the ancestors who gave us life and preserved the land for us and how we, in turn, preserve it for those who will come after us. We are thus compelled to respect this web of life, for all pieces are essential and necessary. Our pattern of living must be one marked by balance—and, at all times, respect for our place as Earth's stewards must be maintained.

The quest for wisdom

Our lives must be defined by a continual search for wisdom and growth. Intertwined with this quest is a need for manifesting the ongoing cultivation of contentment and humility. Happiness is the fruitage of love expressed for others and is not defined by acquiring objects, pleasure, or luxury. Humanity must take from the land over which we are stewards responsibly, with a commitment to taking primarily what is required for our survival. There is room for artisans and nonessential things, but greed is heavily discouraged and usually avoided.

The personal path for each individual

There is a purpose for every living person. It is part of our highest obligation to our family, community, and ecosystem to understand our place in the world. A person must demonstrate their commitment to finding their path and living it to their highest potential. We are not left alone without help in this journey. Any person who sets out on the path of self-discovery will be aided: guides, teachers, and protectors, both seen and unseen, will assist the traveler.

The cycle of change

Everything around us and within us is in a state of constant change. This change occurs in cycles and patterns that are not random or accidental, and we must stay attuned to our own changing needs, both physical and spiritual. This awareness must extend to understanding the needs of our family, community, and ecosystem, which are also changing constantly. Embracing this principle forces us to cast off the needless distractions that interrupt our attention and connection to that which is most important: our stewardship of our life, our health, our family, our community, and our ecosystem.

The bond between the physical and spiritual worlds

The physical world is real. The spiritual world is also real, and they are but two aspects of one reality. Different laws govern each, but they are forever interconnected, and that bond cannot be broken. The breaking of a spiritual principle will affect the physical world and vice versa. For example, if a person is unkind or malevolent, our physical health will be affected adversely. A balanced life is one that honors both, and our actions, thoughts, and intentions must be guarded to respect that sacred bond.

The awareness of our ancestral bond

We are the combined culminating product of our ancestors. Our existence is the ineffaceable consequence of the survival of generations of our ancestors who lived before us. This awareness must be ever present in our hearts and minds, as both an impetus for humility and a galvanizing strength in times of difficulty. Our gifts are not our own but are the fruitage of generations that have come before us, for our genetic inheritance is solely the product of their survival. We gain strength and courage as we bask in the awareness of the obstacles our ancestors faced and overcame. Insight from their multitude of successes and errors grants us wisdom and understanding, and this awareness reminds us of our obligation to our children, grandchildren, and the future generations who will inherit our land. The strength of our successes and the consequences of our failures will live on during their lifetimes, and we are wise to give considerable thought to decisions that will have long-lasting effects.

The four dimensions of true learning

A person must learn in a whole and balanced manner, incorporating the mental, spiritual, physical, and emotional dimensions of their being. The decisions we make affect us on all these levels. Our learning should impel us to care for every aspect of our being, with the ultimate goal of becoming more moral and benevolent beings.

It is an indignation of profound gravity that humanity's history of colonialism, exploitation, theft, subjugation, enslavement, indoctrination, and genocide of countless Indigenous peoples is often presented dispassionately at best and most often simply glanced over in the education of our children. An honest assessment of Indigenous principles reveals the depth, beauty, and value of these cultures and the inestimable value their philosophies and teachings offer humanity. Examining the profundity, beauty, and moral depth of these principles, in contrast with the gross moral deficiencies, greed, and material inequality of the modern world, reveals why these cultures were able to survive for thousands of years.

In less than three hundred years since the Industrial Revolution, the health of our climate and complex ecosystems is declining to the brink of catastrophic collapse. Countless species are actively experiencing extinction, thousands of acres of rainforest are burning a day, precipitous depletion of arctic glaciers and freshwater threaten large segments of humanity. The eventual threatening rise of sea levels poses inevitable risks to large population centers in coastal cities worldwide. The failure to adopt principles that characterized cultures ignorantly deemed inferior has resulted in environmental decay of disastrous proportion. This reality illustrates yet another way the prodigious destructive ability of the illusion of superiority can affect catastrophic consequences on the macroscopic level.

The existence of our present environmental conditions reveals the truth that, as a species, our current path of consumerism, greed, and failed stewardship of the natural world is unsustainable. We must change our behavior as a species, and that behavioral change must be rooted in a shift in the way we educate our children. The change that is needed must be established on the foundation of sustainable principles, and the same principles that allowed Indigenous peoples to live sustainably within diverse ecosystems for thousands of years remain applicable today. Rainforests, plains, islands, mountainous regions, deserts, forests, and others have all sustained various Indigenous peoples. The integration of their philosophies and knowledge of their specific ecosystems sets a starting point for how modern technological advancements can be integrated sustainably throughout humanity.

Some think advancements in technology and sustainable living are mutually exclusive, that the two cannot exist together. But there is no truth in thinking that the two cannot coexist or should not coexist. The reality is that humanity must find the path that allows them to coexist, for our advancement and survival as a species cannot be achieved without the harmonious coexistence of the two. Our population has grown too large to rely solely on antiquated agricultural technological methods, waste disposal, transportation, medical treatment, and a host of other essential services and conveniences that technological advancements provide.

The Value of Money

Only when the last tree has been cut down, the last fish been caught, and the last stream poisoned will we realize we cannot eat money.

—Cree Indian Prophecy

Who doesn't like nice things? A spacious home, beautiful, comfortable and stylish clothes, and a reliable and luxurious vehicle are not things that are wrong in themselves. The illusion of superiority is often rooted in the belief that the acquisition of material wealth inevitably leads to happiness and superiority over others. Human society is greatly enslaved to this fallacy. The truth is genuine happiness has nothing at all to do with material position. While one needs a certain degree of material wealth to obtain shelter, clothing, health care, and basic necessities for living, constant striving for material advantage beyond a certain level of wealth contributes little to lasting happiness and satisfaction. As the well-known proverb states, "give me neither poverty nor wealth."

True happiness comes from connectivity, feeling closeness with others, experiencing mutual respect, solidarity, and a purposeful contribution to the well-being of others. Giving of oneself—sharing friendships and love, and finding one's purpose in service to others and humanity—leads to lasting contentment. Genuine happiness, rooted in an internal condition of the heart, is timeless and indefinitely unaffected by the circumstances of life and can endure even in old age and great material and other loss. While money is essential to sustaining oneself and family, the pursuit of such in an unbalanced way often leads to great misery and suffering. The excessive pursuit of wealth can often lead to frustration, especially if success is delayed; neglect of relationships; and abuse of one's health and psychological well-being. Individuals often place their personal value on how much money they make, or the objects they own, leading again to comparison and feelings of superiority or inferiority to others and renewing the cycle of unhappiness and loss of connectivity. This is not to say that individuals who gain wealth are necessarily set to experience loss of connectivity and unhappiness, but simply that placing the pursuit of such at the expense of relationships and service to others will lead to unhappiness, as does developing an inflated sense of self-worth because of material wealth.

One of the great tragedies of life is to watch individuals spend years of their lives in pursuit of dreams and goals only to find happiness elude them after they've reached them. As a physician, this is a reality I've observed among my colleagues, and even experienced briefly myself. As an intern working 100-to-120-hour weeks, I quickly became exhausted and bored with the rudimentary and repetitive job of performing physicals on patients admitted to the hospital. My weeks were spent with two nights of performing eight to twelve physicals a night and addressing everything from minor to serious complaints of patients already in the hospital. The sheer physical demands of my work hours and responsibilities, combined with the assembly-line repetitiveness of my internship obligations, caused me to lose joy in the profession I'd dreamed of and worked toward since my early childhood. I experienced an epiphany one night and never lost satisfaction in medicine again. I realized I'd lost focus on the true essence of medicine and of being a physician.

To serve others, to care for them in their illness, and to provide comfort and healing to their suffering was what had attracted me to medicine as a child. Watching a sick grandmother conditioned me to be sensitive to the sicknesses and frailties of others. I'd lost the awareness that practicing medicine was not about my own convenience and that to practice medicine sincerely and acquire the factual competence to do so requires ongoing self-sacrifice. Medicine, practiced at the highest level, was, and is, always about the patient and about my desire to serve them while intellectually nurturing my own love for science and medicine. It was an epiphany that helped me regain joy in the career I'd pursued since my early childhood, and more importantly, I realized a truth extending far beyond medicine: *Happiness is rooted in service.* Finding our natural abilities and interests and feeding them with curiosity and passion in pursuit of excellence, with service to others as the goal, is the noblest form of living one can attain.

Without service to others, the monotony of working and making a living can easily become emotionally draining and can rob any person of joy. Even jobs that are monotonous and intellectually or personally unfulfilling can take on greater meaning when we realize the full extent of what our work accomplishes and the interconnectivity we share with all humanity. For example, work that helps to provide an honorable service to others and to provide for our family or meet our individual material needs is an honorable thing even if looked down upon by others. Finding joy in little things can enhance such work. Knowing we are performing a job well is one of the greatest sources of satisfaction a person can gain in life and is certainly better than performing work that is dishonest or injurious in some way to ourselves or others. There is joy in honest work, and deep satisfaction is living honorably.

      This echoes a reality that is often found hand and hand with the unbalanced pursuit of wealth, when individuals engage in self-destructive, relationship-destructive, or community-destructive behavior in their pursuit of riches. The fallacy of wealth's ability to lead to happiness is so powerful, individuals will pursue activities in stark resistance to the prodding of their conscience and the dictates of good sense. Far better to follow the principles that have guided Indigenous peoples for generations, where the constant commitment to honorably living sustainably with nature, and with attention given primarily to the health of one's family, remains the priority.

Chapter Five
The Failure of Education and the Fallacy of Cultural Superiority

I am disturbed, I am uneasy about man because we have no guarantee that when we train a man's mind, we will train his heart; no guarantee that when we increase a man's knowledge, we will increase his goodness. There is no necessary correlation between knowledge and goodness.

—Benjamin E. Mays

Our greatest natural resource is the minds of our children.

—Walt Disney

Of all the great heroes of science who've ever influenced me, and there have been a great many both living and dead, there's never been a scientist for whom I've felt more admiration than George Washington Carver. His brilliance, matched by a magnanimous devotion to education and humankind's welfare, represents the quintessential model for noble education used for society's benefit.

Carver was born a slave. He was a sickly child, a disadvantaged beginning that spared him from the hardship of slave labor and changed the course of his life and human history. His illness gave him time to consider other pursuits, including the study of plants. Though he couldn't read and knew nothing formally of botany as a child, his natural gifts served him well. His inherent affinity for agricultural science eventually led his neighbors to bring him their sick plants, as he quickly garnered a reputation as "the plant doctor," nursing many of their failing vegetation back to health.

When Carver was old enough, he would journey persistently in pursuit of quality education. When he was denied acceptance from one college for merely being a Negro, the term used for African Americans then, he journeyed to another college to continue his education. He eventually earned a master's degree and became the chair of agricultural science at Tuskegee Institute. Carver felt a kinship for poor Southern farmers, and his work on crop rotation helped Southern farmers whose soil grew depleted from the constant growing of cotton replenish their land through the use of enriching soil crops like yams and peanuts.

Carver was most known for his work with peanuts, though he didn't consider that his most significant accomplishment. He developed over three hundred products from peanuts, for example, peanut flour, glue, lubricants, and other diverse applications for personal and industrial use. But he applied for only a handful of patents. He chose instead to share his work in annual publications so that humanity could benefit from it. Thomas Edison recognized his genius, as did Mahatma Gandhi and even the industrialist Henry Ford. Edison offered him a large salary to come to work with him. Carver refused, choosing instead to remain in his humble lab in Tuskegee, pursuing his passion quietly with the commitment he'd sustained from his childhood. He died with a modest amount of wealth but unquestionably with an awareness that he had been exceptionally benevolent toward humanity. His example lives on for the world to see, for he is one of the few truly brilliant and benevolent men in history who placed humanity's welfare above his own profit, despite possessing the brilliance and ability to pursue a different course.

The world is in desperate need of a revolution—not a revolution born of violence but a revolution of the nobility of character. Our planet is sick, the fruitage of generations of humanity living without the proper wisdom and commitment needed to maintain the delicate harmony between our species and the natural world. Among our species' many failures is a failure of education, for proper education is the foundation upon which all things of enduring value are built.

My wife was born in Cuba. She understands and hates socialism and communism in a way I never could. While I always aim to understand complex things in as nuanced a manner as possible, socialism and communism have always been just political theories for me. I never had any personal experience of living under them like my wife did growing up. Her schooling was a blend of traditional instruction in reading and arithmetic, combined with education with the explicit purpose of instilling within her the values of socialism, or, in other words, indoctrination. The books she was allowed to read were stipulated and aggressively censored. Elected writings and poems of Jose Martí, and Fidel Castro's and Che Guevara's life histories were mandated to be read and memorized. She could not progress beyond the first grade without memorizing the story of Fidel's life. When she grew old enough to understand what she'd experienced as a child and the social conditions and poverty she lived through, she understood the value of liberty and freedom in a way only a person who's lived without them ever could.

In America, we don't indoctrinate our children in the way she was in Cuba. Americans indoctrinate our children by what we teach them to value, both by word and example. It is also the absence of other essential teachings that mars American and other Westernized children's education. We focus on educating solely a child's mind, and our world displays the lack of awareness and compassion missing in the very educating of our children—and our environment and humanity have suffered dramatically because of it.

Virtually every public school system in the United States offers programs for talented and gifted students. Nearly without exception, these programs are offered to students who show exceptional science, math, and reading abilities. These students are pushed toward a curriculum that embraces science and technology as the most important of all subjects to be learned, and by their exclusion, students who lack proficiency in these areas are often viewed as less valuable and left to pursue other areas of study. What these curriculums exclude illustrates the most significant manner in which Westernized students are deprived of the quality of education that is most essential.

In 1983, Howard Gardener, an American developmental psychologist, described nine types of intelligence.

1. **Linguistic intelligence:** not just a knack for languages but the ability to use language in an extraordinary way. Great writers, poets, orators, comedians, rappers, and others with this gift can use language with a skill and fluency others simply cannot.

2. **Existential intelligence:** philosophers. One definition of existential intelligence is the ability to use intuition and thought to ask and answer deep questions about human existence. These individuals are often deep thinkers and desire solitude but also feel a deep connection with others and compassion for the suffering around them.

3. **Bodily kinesthetic intelligence:** athletes, sculptors, surgeons, martial artists, gymnasts, dancers, and others who can move their bodies in complex and controlled ways. There is also a sense of timing and mind–body coordination that accompanies this ability.

4. **Musical intelligence:** Musical intelligence manifests as the ability to understand rhythm and tone. Individuals with this type of intelligence are often good at recognizing patterns. They seek out sound and often are skilled at musical

performance and dance. Some can memorize songs quickly and play music by ear and/or recognize specific musical notes by ear.

5. **Naturalist intelligence:** exemplified by people like Steve Irwin, Jane Goodall, and George Washington Carver. It's marked by the ability to recognize and distinguish between various plant and animal life and classify cloud and rock formations. There's frequently a unique connection with the natural world these individuals feel.

6. **Interpersonal intelligence:** the charmers. We've all known them; some people are just likable. Interpersonal intelligence is often associated with politicians, social media stars, teachers, and religious leaders. They understand both nonverbal and verbal communication, are sensitive to both the mood and temperaments of others, and have a knack of knowing how to be likable.

7. **Intrapersonal intelligence:** Know thyself. Individuals with intrapersonal intelligence understand themselves profoundly and can use that knowledge in planning and directing their lives. They are often deep thinkers and shy. Psychologists, religious leaders, and philosophers often manifest this intelligence.

8. **Spatial intelligence:** The ability to think in three dimensions. Architects, sculptors, pilots, and

people with a heightened sense of direction all exhibit this intelligence. Children who gravitate to puzzles and mazes are usually endowed with this intelligence.

9. **Logical–mathematical intelligence:** This intelligence is characterized by using inductive and deductive reasoning, formulating hypotheses, and completing complex mathematical calculations. Scientists, lawyers, mathematicians, detectives, philosophers, and others manifest this intelligence.

There is some debate whether these categorizations are actually types of intelligence or talents. To me, it doesn't matter, really. Humanity often spends too much effort on categorizing things that are obvious. These intelligence types are manifested in history's geniuses in various ways, often with geniuses exemplifying extraordinary ability in more than one of these areas. Our educational system is woefully inadequate at accommodating the diversity of talents and skills present within humanity. We have a system that focuses extensively on technical and external technological development, and our society reflects the emptiness of ignoring the more essential aspects of humanity's needs required to achieve harmonious sustained coexistence with our planet and with each other.

When we consider a number of history's giants like Jane Goodall, Helen Keller, Albert Einstein, Pablo Picasso, Prince, Mozart, Henry Ford, George Washington

Carver, Bruce Lee, Al Pacino, Josephine Baker, Maya Angelou, Bill Gates, Steve Jobs, Shakespeare, Martin Scorsese, Abraham Lincoln, Denzel Washington, Alexander the Great, Socrates, Misty Copeland, and countless others, we see the manifestation of genius in enormously diversified ways. As a society, one of our educational system's most significant failures is that it fails to appropriately value and identify these various intelligence types and gifts in children. We often fail to provide an opportunity in an equitable way for all of humanity and, particularly, disadvantaged communities of the impoverished and people of color. If we are to achieve our optimal potential as a species, our educational system must adopt and adapt new assessment tools with the flexibility to address areas in which society as a whole has been grossly deficient.

Two Unrecognized Intelligence Types

There are two significant types of intelligence that Dr. Gardener failed to present in his groundbreaking work. These two intelligences deserve to be chief among our educational interests, for their deficiency as educational priorities in our society contributes to the degree of moral, ethical, and environmental toxicity that permeates our world.

Empathic intelligence

Throughout history, there have always been individuals who exemplified moral and ethical genius, what I term *empathic intelligence*. They include people like Mother

Teresa, Jesus, Gandhi, Martin Luther King Jr., Malcolm X, Buddha, Mohammad, Harriet Tubman, the Apostle Paul, Jane Goodall, and others whose genius was not only their concern and compassion for others and/or for nature but also their willingness to suffer every manner of injustice and embrace personal sacrifice, even death, for the welfare and aspirations of others. There's a moral courage that accompanies this empathy, and these individuals often are compelled to organize, demonstrate overtly, and sustain through persistent activism a fight for the rights and freedoms of others.

Nature is not without a sense of balance and harmony. And as the world has known its complement of psychopaths and sociopaths, individuals without bounds in their lack of empathy and concern for others, so too nature has given us individuals on the opposite end of that spectrum, uniquely and superiorly empathic individuals. Psychopaths are a blend of genetic and social-environmental factors with a willingness to commit all manner of torture, crime, and murder; empathic geniuses, those who manifest the opposite extreme of empathic intelligence, are also likely to be a blend of these same factors.

As a tree grows in silence, and only makes significant sound when it falls, so too, the identification of psychopaths is made easier by the destruction they leave in their wake, while benevolent lights among humankind often go unnoticed, as is often the nature of those who live humble, self-sacrificing lives. Is it possible for our educational and health care systems to develop mechanisms to identify and encourage empathic characteristics in our populace? Clearly, such an endeavor would be beneficial and worth researching and considering. As genetic analysis becomes more sophisticated, it may help us identify not only physical traits and characteristics but also behavioral traits. DNA analysis, brain imaging, and other studies of psychopaths may indeed help identify the genetic patterns associated with this neuropathology. As physiological and neurohormonal systems often work through cycles and counterbalances, empathy-contributing genes and neurohormonal cycles, and brain morphology patterns relating to empathic genius might well be identified through the study of pathological neuroses.

Intuitive intelligence

Intuition is a real ability and a difficult concept for some people to comprehend. It is particularly challenging for scientifically inclined individuals to accept intuition as a reality. There is this pervasive ignorance that, simply because abilities some describe as clairvoyant or psychic are not experienced by everyone, one who acknowledges clairvoyance as an actuality is somehow irrational or delusional. Talents are not distributed uniformly, as we don't expect exceptional artistic ability in every person we meet; that in no way prohibits the reality that artistic geniuses do exist. Anyone who's had any interaction with a genuine shaman, or any another individual who interacts with unseen entities, knows that there are indeed uniquely gifted individuals in humanity who display clairvoyance.

For some, especially shamans, there is a common ability to see unseen entities and converse and interact with them in powerful ways. Many individuals with intuitive abilities can sense others' emotions and perceive detailed aspects of another's character without knowing them. They may be aware of events that will happen in the future or aware of activities happening in the present from which they are far removed. This ability is documented frequently in writings from various cultures, though little effort has ever been made to characterize it scientifically because such accounts have routinely been flippantly dismissed as erroneous. For example, in any medical textbook that describes abdominal aortic aneurysm symptoms, it will describe that these patients often present to the emergency room with a feeling of impending doom, a sense that something life-threatening is happening inside them, often without any other symptoms. Similar phenomena are documented throughout the world. I've personally observed similar episodes various times in my life, both my own and those of patients and others.

When I was a seventeen-year-old freshman in college, one of my friends developed a severe case of pneumonia. He hadn't eaten in a couple of days by the time I saw him, and I immediately knew he needed to go to the hospital. The way his face looked, a combination of obvious pain and severe malaise, is a look I've seen countless times in emergency room patients. It's a look that's difficult to describe fully with words, but it's something that cannot be faked, something you just know when you see it, especially when you're used to caring for sick people.

His roommate helped me carry him to his car, and we drove him straight to the hospital. The doctors were unsure of his diagnosis initially, and a spinal tap was ordered as part of his evaluation. His twin brother, who was living in New York at the time, some 875 miles away from where we were at Grady Hospital in Atlanta, reported feeling severe pain in his lower back around the same time his brother underwent the spinal tap. The pain dropped him to his knees while he was in the shower, and he described having an accompanying feeling through which he immediately knew something was wrong with his brother.

This experience is not unique. Many twins report having a similar experience when their sibling faced grave illness or pain. Some mothers report sensing their child was suffering or had died while away from them. How can these things be explained scientifically? Some would say that these twins and mothers share a psychic bond with their siblings and children, respectively, in layman's terms, but such experiences are challenging to describe scientifically.

How can we rationally explain such phenomena? This is not magic. There is a reason individuals experience such things. What does it teach us about the possible forces, energies, and bonds that exist around us and between individuals? Can these bonds and forces be identified and understood? What might the technological ramifications of understanding such mean for humanity?

What do these phenomena tell us about the unknown capabilities of the human mind?

Why is it that nearly every person notes having experienced a feeling that someone is staring at them while their back's turned? Essentially, such experiences intimate humans have the ability to sense the intentions of others, at least to some degree. Viewed from this perspective, it's neither illogical nor delusional to anticipate that some individuals among humanity possibly have a more developed awareness akin to this phenomenon. These experiences suggest that there is a force generated by the intentions of others, a force that can be perceived by the human brain, some more than others, and a scientific explanation, not some magical one, why individuals are able to feel when others stare at them.

One possible explanation rests upon viewing the work of Masaru Emoto in a different light. Emoto allegedly did experiments where he obtained microscopic images of water after people focused various emotions at the water. In results heavily scrutinized and doubted, the microscopic appearance of water changed to various shapes depending on the emotions focused. While far from work without its doubters, if Emoto's assertions are in fact correct, could it also be possible focused attention causes changes in the cerebrospinal fluid around the brain? If so, it would provide a scientific basis for understanding in a scientific way how individuals are able to perceive the intentions of others.

When the events occurred with my friend in college, I acknowledged what his brother described feeling is likely to have happened. I didn't know him, and

I see no motive for lying. Back then, I had no interest in understanding what he'd experienced or testing why he'd experienced it. It was just another testimony about something that was beyond my scientific ability to explain as a teenager. Now I see such occurrences as yet another frontier for study, an untapped avenue for essential scientific knowledge to advance in. Many others have described experiencing similar happenings. These types of phenomena must be examined and understood. I believe they hold the keys to understanding what I consider a novel frontier for technological advancement humanity's failed to fully grasp. What some humans can do intuitively and through interaction with unseen entities represents a form of technology of the highest magnitude. Such has yet to be thoroughly evaluated in an organized and disciplined way for the potential and the danger it merits.

In describing and pursuing an understanding of the unknown, words like *magic*, *paranormal*, *occult*, *witchcraft*, and *psychic* are often employed. In my opinion, these are archaic terms, and yet concealed within them lies profound truths that have remained hidden from most of humankind. How is science to understand these phenomena? Are we to hide our eyes and minds in denial of the reality of these and other realities simply because we cannot explain them?

Individuals often ignore such realities because they are linked to what some consider evil forces; for others, these things are simply beyond the realm of possibility. But these occurrences are real. For me, the right questions to ask are, What are the forces behind such events? How might understanding how unseen entities work through shamans and others shed light on forces that exist in the universe that have remained hidden to the vast majority of humankind? Removing the shroud of ignorant denial, religious dogma, and mysticism will lead to a far greater understanding of the unseen forces and unseen entities that operate around us. What is the value that such knowledge may reveal?

For believers in Abrahamic religious traditions particularly, the idea of scientifically studying shamans' work with unseen entities might reflexively be rejected and perceived as deeply offensive. But, I ask, what is the moral objection of seeking to scientifically understand works that will be done whether or not they are studied? Individuals practice shamanism, witchcraft, and other practices, frequently in secret, and no manner of scientific denial or religious opposition will prevent such practices.

I did a fellowship in addiction medicine when I was a medical student. I was appalled after initially learning a famous and prestigious university in New York City was providing crack cocaine to addicts to better understand the drug's effect on the human brain. Initially, I couldn't understand how such research could be justified ethically. As I grew in my understanding of the study design, I understood exactly why the university ethics board approved the study. The addicts were selected only after they had assured the researchers that they had no intention of attempting to stop using the drug. A pure form of crack was provided that was far less dangerous for consumption than the street version, given that any number of other substances often added to cut street crack were not used, and the study participants were to be monitored continuously. The participants' brains were then studied for the benefit of society at large. The university provided the drug to individuals who would use crack anyway, with these individuals' full knowledge and consent.

For those from Abrahamic religious traditions, the thought that unseen entities are a form of alien life might seem revolting. For them, spirits are angels or demons, and, as some believe, only demons interact with shamans. What such research might reveal will have significant religious implications. If the investigation of unseen entities somehow shows that these life-forms do not fit in with the description of unseen entities found in the Bible, Torah, Quran, and other religious texts, religion will experience a significant blow, a crisis of faith. Could it be that what these sacred texts described as angels and demons are simply a form of alien life that's been completely misunderstood? Such would be a crushing revelation to their religious faith. But if such a study reveals that these holy books accurately predicted the existence of unseen entities centuries before modern science revealed their presence, what might the ramifications of that be for religion and the authenticity of these holy books, at least to some extent?

Science faces the same crisis from the study of unseen entities. In the modern world, science has ignored nearly entirely even considering the possibility of unseen entities, falling prey to a type of atheistic mysticism fueled by antireligious sentiment, displaying an ignorance similar to what it often vigorously rejects in religion. How can any human confidently state that they understand everything unseen in the universe or exclude the possibility of life existing in a form separate from organic, carbon-based life-forms? There is an enormous price to be paid for science, too, should the verification of the existence of unseen entities become something that is ever pursued with the intensity it merits. If science proves the existence of unseen entities, what then follows? Questions remain about their origin, how old they are, their intentions, and countless other vital considerations. These realities have tremendous implications. Such a study might very well show that these beings have no connection whatsoever with the heavenly story that proliferates in the Bible and other holy books. But what if they do have some relationship with some form of God or creator? Science, too, will be faced with a crisis, as the bedrock of many influential scientific minds is that there is no such thing as God.

While it's not a futile course to pursue such cogitation as an academic exercise, the real challenge is not to let biases interrupt that which should take precedent. First, the scientific work is done, and then the religious ramifications can be fully considered. Whatever the results that come from undertaking the study of unseen entities, pursuing such a course without biases, with the conviction that having a more accurate understanding of the unseen entities surrounding us will ultimately lead to a clearer understanding of the nature of realities previously poorly grasped. We must accept the truth as it is, not as we wish it to be, and, whatever comes, may our pursuit always be tempered by a search for a greater understanding of the universe and all that comprises it around us.

I find myself in the unenviable position of being potentially adversarial to both bedrock scientific and religious thought. In many ways, the explicit antagonism of my position convinces me that what I am examining must have considerable importance. Finding oneself at the intersection of alienating fundamental paradigms of belief often is the path to great discovery. If the study of shamans and unseen entities yields knowledge that these entities are real beyond question, science and humanity have taken an enormous step in understanding the nature of the universe and answering vital questions regarding the origin of the universe's unseen realities. Invisible entities must have an origin. Whether or not it's related or unrelated to some form of a divine creator, the source of their origin gives us insight into one of the most consequential questions humankind has pondered. The conclusion that will inevitably follow, given that unseen entities are proven real, is that humanity will have made the most extraordinary step in its history in proving God's nonexistence or existence.

It will be beyond difficult for some to reconcile their Abrahamic religious faith with the practices of shamans and others who interact with unseen entities. For some, shamanism is a practice of complete moral revulsion. But there are a great many questions that understanding the world around us and all its hidden realities (beings) will answer. Forces and beings that surround humankind are inadequately understood by humanity in general. Some of these beings influence humanity for good and moral behavior, while others for evil, violent, and devious behavior. At times, even the same entity can influence an individual toward both good and harmful behavior.

How individuals are affected by the evil around them and evil directed to harm them specifically is something people should understand and have defense against. One *espiritista* told me she believes that the individuals who interact with unseen entities to harm led to the existence of individuals who interact with similar entities for protection and healing. The reality is that some individuals use their ability to communicate with unseen entities benevolently, solely for the service of humanity. How are we to think of them? Are they worthy of being considered outcasts, many whose acceptance of their ability was coerced over years of harassment? Is an individual who lacked the freedom to accept or reject something they were born with someone we should condemn? That is an ethical question worthy of the most profound consideration. Society must confront that if we are to work in harmony with shamans and others who interact with unseen entities in an effort to fully understand these entities. To deny the existence of unseen entities and shamans' abilities, and the intelligence and power that unseen entities manifest, is to deny a reality that exists around us.

In my concept of God and the universe, I believe that understanding the true nature of the universe and all it contains is a path to inestimable scientific knowledge. To love is to live in harmony with all that surrounds us, to live with an awareness of all that both exists and is real. To embrace that reality, its good and evil, its ugly and its beauty, as that which it is and not what we hope it to be is to stop hiding our eyes from the truth that is. The reality of all that surrounds us, things hidden and beautiful, evil and ugly, and understanding its ramifications for all beings and the environment around us, moves us a step closer to what some might consider the divine within us. When we live choosing the better aspects of who we are and the gifts we have, the purer side of man's nature and consciousness, is to use all we are in humanity's service. It is neither loving nor reasonable to hide our eyes from that which is true, though it might be considered evil or ugly. Understanding the unknown and the poorly understood and using that knowledge to reaffirm our commitment to what is right is a beauty that dignifies our hearts. To know what we are, what we're capable of, and humanity's place in the universe is hidden in the knowledge, understanding, and deeper awareness of all that remains unknown around us, yet to be discovered and fully understood.

Cultural Superiority

When considering the reality of individuals with intuitive intelligence, I asked myself a question: Why is it that the Western world so vehemently denies the existence of unseen entities when other nations, particularly African countries and other nations with larger populations of Indigenous peoples, regularly acknowledge their existence? If, for example, you ask most Native Americans, Cubans, Africans, Aboriginal Australians, or persons of

Inuit descent if unseen entities exist, they'll reply immediately, as if acknowledging a matter of obvious truth, that such entities do exist. This duality engenders a significant inquiry: why such disparity in thought?

    Often Americans, particularly certain ethnic groups within America, think of racism as only a thing of the past. There's often this notion that racism ended with the end of slavery or the first Black president's election. Racism, and the powerful effect of prejudice, can be observed throughout history—not just in America but in every part of the globe. It has produced tangible and long-lasting effects and in many ways has shaped the course of history and the paths of nations for hundreds of years. The impacts of colonialism, the Crusades, Indigenous people's religious indoctrination, slavery, apartheid, India's caste system, and other forms of prejudice and bigotry still linger today. Its consequences have crippled humanity's ability to achieve its optimal potential. The denial of the existence of unseen entities reflects the subtle but powerful effect of prejudice. Every Indigenous culture that has ever existed, including those still present today, describes these entities' existence. Yet Western civilization has clung to the belief that these numerous peoples are pitifully ignorant, enslaved to some sort of mass delusion. The very refusal of society to accept the innumerable accounts of these predominantly non-White peoples displays how prejudice has the power to supplant the optimal destiny of humanity. Had such accounts of these entities come predominantly from nations and peoples of European descent, they would've long been met with the assumption of accuracy, and the subsequent study of these entities would long ago have changed the course of human history.

Cultural superiority is an ongoing problem within the educational framework of America. I cannot speak of other nations, but I believe something similar persists throughout the world, as racism and prejudice are as alive today as they've always been and the wave of right-wing political movements around the globe clearly displays its ongoing, not so indolent, presence. The thinking of people is often reflected in the literature they espouse, and educators are no different. When I was in high school, I was mandated to read books from writers whose perspectives and fictional narratives were nearly entirely White and Eurocentric.

My entire high school curriculum of four years of English, in a school more than 50 percent Black, involved not a single novel, poem, or nonfiction book written by a person who was non-White. Instead, I was inundated with books like *The Great Gatsby*; *Huckleberry Finn*, and its prolific use of the word *nigger*; *To Kill a Mockingbird*; Edgar Allen Poe; the Puritan; Planter; the Pilgrim Period; and the like. Whether this was an intentional oversight or just the ignorance of racially aloof educators, I cannot say. In thinking back, my previous statement was incorrect. I had to read one book by an African American writer, Richard Wright's *Native Son*, a story about a poor Black boy who accidentally murders a White woman.

Works by Shakespeare and Sophocles, pieces like *The Iliad* and *The Odyssey*, and other established masterpieces are appropriate works to be enjoyed and studied. Still, they are by their very nature antiquated and racially extremely homogenous. The world is full of brilliant writers, artists, musicians, and other geniuses whose works and ethnicities are as varied as the diversity of their genius. By what more significant gesture can we demonstrate what and who we value as a society than by the manner in which we educate our children? By the very nature of who and what we exclude from our education, we also demonstrate our society's values. For years, a segment of educators and others have complained about the cultural biases within standardized tests like the SAT and ACT. These tests have their failings. The real problem is that the system of education exudes cultural biases at every level and that these tests fail to evaluate students in the areas that matter most.

How do you measure a person's integrity, honesty, empathy, leadership ability, determination, ability to overcome obstacles, creativity, work ethic, spatial intelligence, commitment to environmental conservation, interpersonal skills, and other qualities essential in the modern world and beneficial for our society? When we consider these vital qualities and abilities and look at how we evaluate students, it is clear such antiquated instruments that focus principally on factual knowledge and reading comprehension are ineffectual. We have completely ignored other factors that are far more critical indices of character and more effective in predicting a person's likelihood for success. These instruments are the tools of an antiquated educational system that has failed to instill in humanity the qualities needed to maintain a harmonious balance within our society and with the natural world. For as Mays thoughtfully illustrates, "we have no guarantee that when we train a man's mind, we will train his heart; no guarantee that when we increase a man's knowledge, we will increase his goodness. There is no necessary correlation between knowledge and goodness."

The Void of Moral Instruction

When we take a closer look at our educational system, the entirety of its focus, except for, say, religiously inclined private schools, is on instilling factual knowledge and reading comprehension. Simply put, we focus on the mind and not the heart. This approach is reflected in the macrocosm of society, a society that focuses on success defined and measured by individual wealth and educational attainment, essentially acquiring things, void of any moral, ethical, or environmental conservational principles. Is there any wonder we see our planet in ecological decay and a world fractured by war, moral decay, and the persistence of gross degrees of inequality? If our children are not educated in the values that are principally needed for sustainable existence with our planet and our species, how could they manifest these values as adults?

In decades past, efforts for moral instruction were made but always interwoven with religious indoctrination, for example, prayer in school and reading the Bible as a form of daily ritual. Parents appropriately withdrew from such efforts, believing that religion is a diverse and personal choice and that the state should not impose religion on the populace. Religiously minded people, particularly Christians, frequently believe only Christians are people with morals, but this is far from the truth.

Practically every religion advocates ethical principles. Even those without belief in God can often manifest ethics and moral behavior of the highest degree. At times, those who present themselves as religiously devout individuals manifest behavior that is far from moral and ethical. It is possible to instill benevolent values within our children as a part of their educational experience, and it must be encouraged with the most extraordinary vigor. Encouraging benevolence, including principles like honesty, integrity, proper work ethic, empathy, generosity, tolerance, impartiality, and other universally recognized beneficial qualities in our children need in no way be connected with religious indoctrination and instruction.

A child's mind is a sponge, and children typically understand even the complex concepts of fairness, equality, impartiality, and benevolence. Watch any child playing, being disciplined, sharing toys, and fairness is a concept they easily comprehend. As prejudice is a learned pattern, so qualities of benevolence must be instilled. It is a failure of considerable consequence to delay such instruction in favor of exclusively limiting education to the mind.

It's easy to leisurely criticize aspects of society without offering any solutions. And while it's challenging to address all the complex problems with society, I have salient thoughts on the path forward to bring more balance and harmony for humanity and our fragile environment. A lack of empathy and awareness has contributed to the development and sustenance of a significant portion of these issues. Still, education is a powerful tool, and it is the only real pathway to remedy the enormous challenges our species and civilization now face. As technologies push society forward and globalization, artificial intelligence, and other aspects of external technological development drive us further away from principles of benevolence and sustainability, the true path forward rests in embracing the philosophies and characteristics of cultures long ignorantly thought primitive. Considering ways the modern world can learn from the wisdom of Indigenous peoples is perhaps the most essential innovation in education a planet on the brink of climate catastrophe could benefit from.

Chapter Six

The Ignorance of Prejudice and the Power of Benevolence

We have learned to fly the air like birds and swim the sea like fish, but we have not learned the simple art of living together as brothers. Our abundance has brought us neither peace of mind nor serenity of spirit.

—Dr. Martin Luther King Jr.

None of us is responsible for the complexion of his skin. This fact of nature offers no clue to the character or quality of the person underneath.

—Marian Anderson

There was a war within the Vietnam War seldom openly acknowledged and discussed. During one of the greatest periods of racial tension in America's history, not surprisingly, the conflict between the races spread onto the battlefield between American soldiers. In addition to being placed more frequently in forward combat roles in Vietnam, Black soldiers were often the victims of brutal violence and killing by White soldiers both on and off the battlefield, and vice versa. Dozens of Black soldiers, perhaps even more, formed a fraternity to protect themselves from the violence they faced from their fellow White US soldiers, and admittance into this fraternity was contingent upon completing a gruesome act. To gain admission, they had to kill a White US soldier.

I've met a few members of this fraternity at various times in my life, but one in particular, during my time photographing in Atlanta. He shared many stories with me, but there was one I think especially worthy of passing along. It was common during the period of the Vietnam War for White MPs to sick their dogs on Black soldiers, both within the states and in Vietnam. It was so frequent that Black soldiers trained on how to defend themselves from dog attacks. Such an attack happened to the veteran who told me of this fraternity, and the training he received from his membership proved useful during one particular episode.

My acquaintance had difficulty with several of his White fellow servicemen. When I met him, he was in his late forties to early fifties. He was tall, six foot four or so, and was a muscular and physically imposing figure. I can only imagine his physically intimidating stature as a young man. He was not the kind of guy who would cower to anyone, and during his time in the military, his confidence and candor were not infrequently poorly received from anyone with bigoted thoughts of how a Black man should conduct himself. His disposition garnered the disapproval of one White MP in particular, and he decided he would test his dog's capability on his fellow soldier.

When the MP released his dog on my acquittance, he ran out of sight of the MP. While running, he removed his belt and made a loop with the square metal buckle so that it had a handle at one end and slip loop at the other. When the dog leaped to attack, he looped the belt around the dog's neck, tightened it to suffocate him, and punched the dog repeatedly in the nose until it whimpered and ran off. He removed the belt before the dog ran back to the MP. Thereafter, when he walked pass the MP, the dog would whimper and hang his head. The MP, realizing my acquittance must have done something to the dog, yelled, "What did you do to my dog, nigger?"

We don't often think of racism as having tangible effects on history, but it does, and its effects have changed the course of history. Colonization, the Holocaust, the Native American genocide, American slavery and the Jim Crow period after, and even the modern-day biases within the criminal justice system all have altered the history of nations. But it's not just the effects of prejudice on the macrocosmic level that have significance. Bigotry and prejudice have catastrophic effects on the lives of people individually, and not infrequently.

It's a multisided dice, however, and every race is not without members whose hatred extends to others without any justification other than the color of their skin. I've never understood that. It's just a mentality that's intuitively flawed and erroneous to me. But what seems obvious and intuitive to me no doubt was significantly affected by the formative examples that molded me as a child. When I was young, my adopted parents set an example that's stayed with me through the entirety of my life. My adopted father, with a near genius level IQ, is still to this day the fairest man I've ever known. Bigotry is a nonstarter for him, and far removed from who he is a person. My mother is just the same, even more so. While my father is an exceptionally quiet man, with never more than a couple of words to rub together unless spoken to and prodded, my mother is the exact opposite. She's a social butterfly, and my mother's large collection of close friends always filled the house on holidays, birthdays, graduations, and the like. Her friends and their families are my extended family, and many of her friends are my "aunts and uncles," and their children, my "cousins."

My mother exudes love, and anyone who knows her knows that love is the essence of her character. At Christmas, she'd ask us to give one of our gifts to the less fortunate, a painful sacrifice for me then, and we'd go to various places donating those gifts to families who couldn't afford them. Never the homebody, my mother's always off on some task, serving her church family or others for whom she has affection. My grandmother lived with us for the entirety of my life, and I watched my mother be her caregiver as her health failed the final ten years of her life or so. Next it was my aunt, in her eighties when she came to live with us, and my mother again played the role of caregiver, first in our home and then in the nursing home as my aunt's health worsened and my mother could no longer manage her care in our home. But she didn't abandon her there. For the rest of my aunt's life, my mother made the trip to the nursing home, some thirty miles one way, several times a week, bringing her food and desserts. She did this for more than twenty years, including my aunt's last seven years in the nursing home until she finally died, and she did it ungrudgingly.

These are the parents who raised me, two people of exceptional character with a marriage rooted in love and fidelity. Few children often see such examples in the parents, and orphans, almost never. My parents are a beautiful and precious gift life granted me, and their love, intellect, and humility rooted principles in me that I most certainly wouldn't have known without the fortune of having them in my life. And they were both without a bigoted bone in their body.

My parents were both from the South, my dad from Republican ruby-red Alabama, and my mom from Tampa. My father's father was a coal miner, and my father, the youngest of ten children. When he was young, from the age of seven or eight, my dad woke up every day at 4 a.m. to start the fire in the family stove with coal that also served as the heater for the small two-bedroom house. He was no stranger to racism in Alabama and often shared stories of the challenges he faced growing up in a community where an exceptionally intelligent Black child was often overlooked in favor of lesser gifted Whites.

When he was twenty, a drunken White man ran him off the road while he was driving. The man swerved and crashed into a tree, badly wrecking his own pickup, but he was uninjured. The police were summoned to the scene, and the man accused my father of forcing him off the road. My father thought it prudent not to accuse a White man of lying to a White police officer in deep-red Alabama in the 1950s, so instead he took the policeman a short walk from the vehicles and pointed to the skid marks that lead from my father's side of the road to where the man's truck had crashed into the tree. The police officer let my father go on his way and arrested the man, realizing likely from the smell of his breath, and the skid marks on the road, that the accident was not my father's fault. That episode could have easily gone a different way, and had the officer been a less honorable man, my father's life could've been changed forever. Unfortunately, countless other minorities and innocents are all too often not so fortunate.

My father wasn't the activist my mother was. His disposition wasn't one to garner attention, he preferred to avoid it. He'd seen random lynchings in Alabama as a child and seen Black soldiers disciplined more severely than his White counterparts during his time in the military. He chose to handle prejudice by avoiding conflict and maintaining a low profile. His entire focus was on securing and keeping a livelihood for his family. My mother was of an altogether different mind-set. She actively participated in sit-ins and other forms of nonviolent protesting during the 1960s civil rights movement. Her mother, the grandmother who lived with us, was a caterer, an exceptional cook who'd seen the effects of prejudice firsthand working as a server in some of the finest restaurants in Washington, DC, in the 1930s through the 1950s. She also shared many of her experiences with me as a young child, and I was an eager budding journalist with a listening ear even then.

One day my grandmother served a wealthy White woman in a restaurant in Capitol Hill. Her frequent customers were congressmen and other affluent clientele. The woman complained that her coffee wasn't hot enough and insulted my grandmother's intelligence and berated her loudly. She took the coffee back the kitchen, boiled it, and then brought it back to the woman. "Is it hot enough now?" my grandmother asked. The woman burned herself while eagerly tasting it the second time, to my grandmother's vindictive intention no doubt.

I'm not sure if it was her mother's experiences or living under the reality of segregation in Tampa and watching Whites treated differently, but my mother deeply resented segregation. When the call for activism grew in the 1950s and '60s, she answered the call eagerly. When she moved to DC, she became a paralegal and worked for the Equal Employment Opportunity Commission (EEOC), while my father worked as one of the first computer scientists in the government as a programmer for the Department of Housing and Urban Development (HUD).

I share these accounts of my parents because they illustrate two completely different approaches to dealing with prejudice, and how they both affected my perspective. My father preferred avoidance of conflict, dealing with threats by being prudent and kind, and letting his intellectual excellence speak for itself. My mother chose a more active role, preferring to fight first through activism and then through legal remedies as part of EEOC.

When I was young, these stories affected me, as did a subtle, more subconscious awareness of the reality of prejudice and racism. I chose science because I loved it, yes, and the human body fascinated me, but I also realized, as did my father, that science and mathematics are among the most objective fields a minority can pursue. Either you can do the equation, or you cannot. Either you can synthesize the organic compound, or you cannot. Science and mathematics are the great equalizers of objective assessment, in stark contrast to the arts and humanities, with perhaps the exception of music, and excellence is far more difficultly denied.

Science and math don't care what color you are. Your ability can be accurately assessed, and even the prejudiced are forced to recognize excellence in a way, bias often prevents artistic talent from being acknowledged. This is not to say that there aren't objective ways to measure ability in the arts. There are. For example, in music, either you know a musical scale, or you don't. Either you can improvise, play by ear, and read music, or you can't. Painting is not so dissimilar when it comes to realism. But fiction writing, poetry, playwriting, sculpture, and abstract art are often very subjective, and as a minority, it can be exceedingly difficult to gain success in a genre where cultural subjectivity dominates what influencers define as beautiful. And as a minority, subject matter that deals with topics you have a more personal experience with, by its very nature, is geared toward a narrower audience than something for White audiences.

When we look at the modern world, large segments of the populace often think of prejudice as a thing of the past, without modern-day ramifications. There's an illusion that the world has progressed to a place where prejudice is insignificant. Because there's been progress, and some would argue not significant progress really, that certainly doesn't mean prejudice is no longer a threat to humanity. This fallacy is ever present in the world, and its existence must be confronted and fought against if our species is ever to embrace our full potential. I firmly believe some of the world's greatest artists and scholars have lived and died without ever being known, simply because they were born in less than supportive circumstances.

When I was eighteen, my mother bought me the most cherished material possession I've ever owned, besides my camera and guitar. I'd come home for Christmas break after my first semester at Morehouse, and my mom gave me a wool coat for my eighteenth birthday. The coat was a light winter coat with a pattern customarily associated with Native Americans found throughout the Southwestern US At the time, I'd never seen the like of it, and I felt a fondness for the coat immediately. There was no one in my school who had a coat like it, and it made me stick out among my peers. I eventually had a vest sown into the coat, so I could wear during the cooler temperatures not uncommon in Atlanta. I kept it for fifteen years or so, until the coat could no longer be patched and holes and tears became too unsightly, even for the rugged careless approach I had for fashion in those years.

Perhaps it was that coat that set me on my journey to understanding the Indigenous. Whatever it was, in time, my studies grew to include Indigenous people's history and culture. When I read the beauty of Indigenous philosophies, and saw places like Machu Pichu, Angkor Wat, Ancient Timbuktu, Ma'at, Teotihuacan, the magic of the Egyptians, and the beauty of the Brahmans, and things like Tibetan monks' physical prowess and the healing of Innuits, Siddhas, and Native Americans, I recognized the beauty of their genius, though in a form distinct from the Western world. The truth of unseen entities was accepted by these cultures, and hidden principally by the power of prejudice. The strength of TOTI, *technology of the indigenous*, holds promise that cannot be quantified, not in small part because it offers guidance regarding the origin of our species and insight into forces and beings that humanity largely remains ignorant of. Imagine how much progress, in every field of human endeavor, has been delayed because of the ignorant and inaccurate characterization of these enlightened peoples simply because their manner of life, their manner of dress and makeup, and the way in which they practiced religion is not as Western civilization perceives it should be.

Lessons from Ancient Cultures

Library of Timbuktu

The library of Timbuktu was the jewel of the ancient Mali civilization. Located at the Niger River on the border of the Sahara, Mali became a thriving civilization in the fourteenth century. The city attracted a large number of scholars and intellectuals, and the library at Timbuktu is considered one of the most significant educational sites of the ancient world. The city also featured a university and a massive mosque, which still exists in the Timbuktu today.

Teotihuacan

Located some thirty miles northeast of modern-day Mexico City, Teotihuacan was the largest city if pre-Aztec central Mexico. With an estimated 200,000 residents at its peak, Teotihuacan was an important commercial and educational hub in the ancient world. Its pyramids are still among the greatest wonders of the ancient world, and the beauty and durability of their design and construction displays the intelligence of the sophisticated civilization that constructed them.

Machu Pichu

At 7,700 feet above sea level, Machu Pichu rests high in the Andes Mountains in Peru. Literally translating to *Old Peak*, Machu Pichu is significant for several reasons, not the least of which is that is represents one the few pre-Columbian ruins left largely intact. Believed to be the site of an ancient palace built in the mid-fifteenth to sixteenth century, the ruins represent an elaborate, intricately connected series of terraces and rooms thought to have functioned as a retreat for royalty and other elites. The city was abandoned for an unknown reason, but lack of water appears to be a possible explanation.

Angkor Wat

Comprising hundreds of temples spread over more than four hundred acres, Angkor Wat is the largest religious construction in history. Its distinct construction style speaks of the great architectural diversity between traditional rectangular Western architectural constructs and more free-form structures found more prevalently in the East.

The Blue Mosque

Europe is routinely seen as a historically rich land, and for good reason. Its collection of castles and Christian edifices offer a unique view of the past through the lens of its rich architecture. But the Muslim world has its parallels, too, and has no shortage of spectacular ancient edifices. The Blue Mosque is one such structure. Built between 1609 and 1616, the world's largest mosque is a colossal architectural marvel, and it gets its name from the blue tiles that line its interior. With enough room to accommodate an estimated 30,000 people, the Blue Mosque is the largest mosque in the world. Its 20,000 handmade ceramic tiles speak of the craftsmanship that went into the seven-year construction project.

Public works of Egypt

We commonly think of the great pyramids at Giza and Egypt's Sphinx when we think of the wonders of the ancient world, but Egypt was home to many more fascinating public works. With a cadre of unique astronomical features planned into the dimensions and locations of these structures, a study of them is richly rewarding and reveals the depth of understanding the Egyptians possessed of both architecture and astronomy.

Each of these structures represents not only an architectural achievement of consequence, but their locations were spiritually significant, educational and/or commercial centers for their cultures. The nationalities that were associated with these structures each had their own unique religious perspectives, diverse economies, and scientific achievements. The diversity of these cultures and the significance of their achievements represent a beautiful avenue of educational emphasis that should be highlighted. Often times when history is taught, we emphasize things: dates, wars, treaties, and the like. History should be more about the world's peoples and their accomplishments, in my opinion. In an era in which racial and political division is as significant in the US as ever before, our efforts to improve race relations should alter the way we teach history. A focus on the architectural masterpieces and philosophies of various cultures offers a unique opportunity to help children appreciate the value and sophistication of cultures often excluded from childhood education, an endeavor as important as any in instilling egalitarian principles in young ones and overcoming racial bigotry.

  The locales I listed previously, and many others, offer unique opportunities for scholastic instruction. First, they're super cool structurally and architecturally. They're beautiful to look at and often have unique and impressive astronomical features that illustrate both the depth of these cultures' engineering and intellectual prowess and their artistic sensibility. Cultures and countries frequently ignorantly perceived as inferior have produced some of the most amazing architectural and philosophical contributions the world has ever seen. It is a disservice to humanity to ignore the value of all of these cultures' contributions to the world. Nothing is as effective at destroying the ignorance of prejudice as educating children properly. Mark Twain wrote,

> Travel is fatal to prejudice, bigotry, and narrow-mindedness, and many of our people need it sorely on

> these accounts. Broad, wholesome, charitable views of men and things cannot be acquired by vegetating in one little corner of the earth all one's lifetime.

Twain saw the power of travel to eradicate the ignorance of prejudice and bigotry. But what about the power of education to achieve the same for our children? Should we expect every child or person, particularly those from disadvantaged circumstances, to depend on travel to help eradicate the indoctrination of prejudice and bigotry? That would be a grossly inadequate approach to addressing the problems of race and prejudice in society. Instead of literally taking people to unique peoples and places, we can far more easily bring those peoples and places to them in the beauty of the books and other tools useful for educating our society.

We've entered a unique time in humanity's technological advancement that allows for the production of extraordinarily powerful educational tools never used before. Tablets and computers allow for the incorporation of 3D models, animation, video and drone footage, photographs, artifacts, didactic text, augmented and virtual reality, and the possibility for other tremendous innovation in educational technology. The production of interactive textbooks and sophisticated computer educational games humanity's never known the full potential of. With the added caveat of being paperless, such texts could be produced and used for everything, from history to mathematics and science. Whoever has the capital and the foresight to develop this next generation of interactive textbooks will lead a new frontier of educational opportunities that will be unparalleled in humanity's history. I only hope such efforts will be driven by those whose understanding of the import of moral and ethical instruction is as developed as their commitment to innovative factual instruction.

The Power of Benevolence

I wish I could call this section of the book "The Power of Love." Because the word *love* has lost so much of its meaning in today's world, I decided I needed a different, less frequently used term. Benevolence feels more connected to the meaning of what the *love* I'll be discussing really is. Principled love is a quality displayed through the sincere thoughtful actions of one person in pursuit of the welfare and happiness of others.

The concept of principled love is not one born from my own originality. It is a term derived from the Greek word *agape*. When I became fascinated with religion and holy books, especially the Bible, I paid considerable attention to how love is depicted and expressed within them. I wanted to understand what love means to various cultures and peoples. Within the Bible, principally within the New Testament, various concepts of types of love are expressed by the use of different Greek words.

Eros—passionate or romantic love, like the kind felt between couples.

Philia—a fondness and camaraderie similar to what is displayed between family and close friends. Twice when

Jesus asked Peter if he loved him, Peter fervently confirmed his love and affection for Jesus, using word *phile*, the form of love associated with brotherly affection.

> 15 When they had finished eating, Jesus said to Simon Peter, "Simon son of John, do you love me more than these?" "Yes, Lord," he said, "you know that I love you." Jesus said, "Feed my lambs."
>
> 16 Again Jesus said, "Simon son of John, do you love me?" He answered, "Yes, Lord, you know that I love you." Jesus said, "Take care of my sheep."
>
> 17 The third time he said to him, "Simon son of John, do you love me?" Peter was hurt because Jesus asked him the third time, "Do you love me?" He said, "Lord, you know all things; you know that I love you." Jesus said, "Feed my sheep."
>
> —John 21:15–17

Philadelphia, a derivative of this word, literally means "affection for a brother."

Agape—principled love, a benevolence that extends to all people. This is the type of love most commonly spoken of in the New Testament.

1 John 4:16 tells us that God is love. The Greek word used in this scripture is *agape*.

> 16 And so we know and rely on the love God has for us. God is love. Whoever lives in love lives in God, and God in them.

This idea of God's predominant nature being linked to *agape* is expressed throughout the Christian Greek scriptures. Other scriptures reinforce this.

> [44] But I tell you, love your enemies and pray for those who persecute you, [45] that you may children of your Father in heaven. He causes his sun to rise on the evil and the good, and sends rain on the righteous and the unrighteous.
> — Matthew 5:44-45

Philia would be both inappropriate and unreasonable to require of Christians in all circumstances, for affection for someone who has injured you significantly, such as through slander, theft, or murder, would be unattainable. But principled love, manifested through not injuring a person in retaliation for example, would be an expression of *agape*. It is this type of love that is so significantly deficient in society.

Principled love makes the possibility of prejudice far less likely and caustic. For such love would prevent individuals from taking actions that would cause harm to others. Racism is like any weaknesses individuals have. For example, a person might have violent thoughts or an inclination to steal, use drugs, or abuse alcohol but choose to avoid such actions for any variety of reasons. Similarly, a person may have biased or bigoted thoughts, but principled love would make behaving in a way to cause injury to another an action that would be unacceptable. It is an impersonal love, one that permeates a person's character and actions, regardless of the individual with whom or the circumstances with which one is presented. Such love, for example, may prevent infidelity for some when a spouse has been unkind or a marriage is under stress. In this case, fidelity may be driven predominantly not by affectionate commitment to one's spouse but out of devotion to personal principle: "I will not cheat because of who I am, not necessarily because I'm still head over heels for my partner or I feel happy with the quality of the marriage." A person may consciously choose to honor their marriage vow, despite the unpleasantness of current circumstances within the marriage, out of personal commitment to keeping their word and hope that the relationship will improve.

The Quran speaks of principled love in similar ways

> There certainly has come to you a messenger from among yourselves. He is concerned by your suffering, anxious for your well-being, and gracious and merciful to the believers.
> —Surah At-Tawbah, 9:128

> Never will you attain the good [reward] until you spend [in the way of Allah] from that which you love. And whatever you spend—indeed, Allah is Knowing of it.
> —Quran 3:92

> And do good; indeed, Allah loves the doers of good.
> —Quran 2:195

Buddhism and principled love

Buddhism's thoughts on principled love are considerably more diverse, and even more intensely comprehensive in some sects, than what is expressed in Christianity and

Islam. For example, some sects within Buddhism seek to protect all life, and even face masks are worn to prevent inadvertently killing insects.

It has often been said that kindness is weakness, charity is foolishness, and mercy is a mixture of them both. The persons who uttered those words, and others similar, failed to understand the tremendous power having a benevolent disposition grants one in life. Let's consider how benevolence benefits one in many aspects of life. While *The Illusion of Superiority* deals primarily with how we see ourselves in relation to others in the world around us, "The Power of Benevolence" deals primarily with how we treat others, how we think about life, and how those actions and thoughts affect us personally and the individuals around us.

Love and benevolence are principally manifested by the conduct they engender. In romantic love, especially when new, affection and kindness spring from the heart naturally, and the warm attachment felt between the pair blossoms. Seasons of trials and difficult circumstances can strain even the strongest of romantic attachments. Often there is this commonly used expression that love isn't enough. That is a ridiculous conclusion, for true love is capable of solving every manner of relationship difficulty, and there is no force stronger, and no energy more powerful, than love.

> Place me like a seal over your heart, like a seal on your arm; for love is as strong as death, its jealousy unyielding as the grave. It burns like blazing fire, like a mighty flame. Many waters cannot quench love; rivers cannot wash it away.

— Song of Solomon 8:6

> [4] Love is patient, love is kind. It does not envy, it does not boast, it is not proud. [5] It does not dishonor others, it is not self-seeking, it is not easily angered, it keeps no record of wrongs. [6] Love does not delight in evil but rejoices with the truth. [7] It always protects, always trusts, always hopes, always perseveres. [8] Love never fails. ...
> — 1 Corinthians 13:4-8

Deep love inspires one to place the interest of their partner ahead of their own personal interests, be it emotionally, physically, or financially. It would inspire a lazy person to work harder to provide for the welfare of their spouse and children physically, overcome addiction, maintain fidelity, and forgive minor, and sometimes even major, grievances like adultery.

A common fallacy people relish is that individuals who display benevolence lack awareness, the sober realization of the dangers and risks in the world around us. There is this commonly embraced ignorance that benevolence simply exposes one to being taken advantage of, and as such, benevolence is weakness. Such a perspective is often born from the fruitage of cynicism, albeit cynicism deeply rooted in personal experience and observation of the selfishness, greed, and viciousness that is common among humankind. But such thinking cannot be further from the truth.

The pursuit of benevolence is not a pursuit made out of ignorance or foolishness. It is a noble endeavor, made soberly and consciously and with the hope of improving the quality of life of the individuals with whom we share a global community. It's done with an awareness that cultivating benevolence, and similar qualities, also improves our own lives, both physically and emotionally, and improves the quality of our relationships both personally and professionally.

This is part of the beauty of both the Bible, the Quran, and other holy books. The depth of the ethical principles they espouse are just as worthy of academic consideration as traditional philosophical schools of thought. The breadth and depth of love as is presented within the context of religious writings is often discounted by intellectuals and others simply because such thoughts are expressed through religious texts. But the expression of such thoughts, and the depth of their philosophical beauty, is not diminished by their connection with religious ideology. Why should these religious texts not be met with the same impartial scrutiny for their intellectual value as any other masterpieces of literature? Instead of discounting religious thought, the elements of religious texts that espouse universal thoughts of benevolence should be embraced and studied, for if their principles of benevolence are incorporated into society, the quality of interpersonal relationships will be improved.

For many, happiness is so elusive because they choose superficial goals, mistakenly believing these will make them happy. Material prosperity, beauty, fame, artistic achievement, educational and professional pursuits, the acquisition of power and influence, and the like are all common goals that in themselves will not lead individuals to happiness. Intrinsically, none of these things are wrong to achieve or acquire; the error most commonly made is that people often set these as their preeminent goal. Ignoring our most important and basic needs for the acquisition of superficial goals that will not lead to true fulfillment is the most certain way to achieve a life that is empty and unsatisfying. To love others, to be active in caring for them and knowing and experiencing love in return, is our most basic and fulfilling need. The more we pursue satisfying this need, the more our endeavors will contribute to our happiness and well-being. Conversely, pursuits that distract us from satisfying our most basic need will ultimately lead us to feeling less fulfilled and ultimately less happy. This is a simple yet profound truth, often ignored or dismissed as trivial. Individuals often spend significant portions of their life pursuing goals they believe will make them happy, only to find that once they've been achieved, they do not produce the satisfaction anticipated.

It is not a trivial accomplishment to understand the value of benevolence and to pursue goals related to satisfying our most important inner needs. This is the accomplishment that matters most, and thinking about the ways we will benefit from this pursuit is worthy of our consideration.

A Choice to Understand Benevolence and Love

I had an epiphany sometime during my junior year in college and made a conscious decision to pursue to the best of my ability understanding what it means to grasp, and practice, principled love. I realized in that moment that pursuit was the most important thing I could ever achieve in life, for it would lead me to genuine happiness and being a benevolent member of my family and community.

At the time, I had no desire to write. I only wanted to understand the world and my place in it. I felt saddened by an observation I made. I saw that when I observed most strangers, there was a deeply entrenched cynicism that often led them to being skeptical of genuine efforts of kindness people extended. The appreciation of that actuality, more than any other, helped me understand the depth of toxicity among humanity, for what does it say about our civilization when genuine acts of kindness are most often met with skepticism and suspicion of ulterior motives?

Living in a civilization where kindness and benevolence are predominantly met with cynicism and skepticism does not encourage the pursuit of those ideals. Yet the pursuit of them is intricately linked to our own well-being, both physically and emotionally, as well as that of society. We don't often think of kindness and benevolence as traits that contribute to success and prosperity. In fact, there is a common notion that kindness and generosity are possessions only of the weak. This ignorant characterization of these qualities, and others like humility, reveals a grossly shallow depth of understanding within society. Anyone who's spent any time in an activity or profession performing menial tasks or serving others knows humility is a tremendous asset. What is more, humility and benevolence are at the heart of achieving genuine success. True success, defined by building and sustaining loving friendships and relationships, serving the needs others, meeting family obligations, showing generosity, displaying principled love, is made easier by humility. Individuals who consider humility dispensable likely perceive success not in terms of connectivity, quality of friendships, relationships, and purposeful living. For many people, in the acquisition of material wealth and the advantages of fame, prestige, power, and the like, often at any cost, a lack of humility and empathy can enhance such pursuits, for such things are not infrequently gained by disregarding the needs, rights, and value of others.

Throughout my life, I've been fortunate enough to have lasting friendships and relationships, including my marriage. These relationships have brought me deep satisfaction, though not without their ongoing challenges and difficulties. There are few things more comforting than having a mate and friends who understand you and with whom you feel free to share your innermost thoughts and frailties. I once heard a quote: "True friends are the ones who know everything about you and still love you." This statement gets at the heart of the openness and vulnerability that rests at the center of genuine relationships. Despite our best efforts, we make mistakes. We say things that are unkind, make decisions that are selfish and short sighted, and make other mistakes that can place strain on our work relationships, friendships, and families. Humility helps us to be forgiving, an essential quality in a healthy and lasting relationship, as we all make mistakes and need forgiveness from time to time.

Feelings of entitlement are at constant play in the world. They affect how we deal with our family, strangers, and workmates and often rob people of joy and peace when they perceive what is justifiably due them is not given. It affects how people react to tragedy and difficulties, and especially our ability to endure difficult circumstances. Instead of focusing on finding resolution to the personal loss and difficulties, an entitled person will be unbalanced with distracted thoughts about the unfairness of the situation, crippled by the "this isn't supposed to happen to people like me" syndrome. Feelings of entitlement distract from coping with and managing the loss in a way a person less inclined toward feelings of self-importance would face them.

People often spend more energy trying to convince someone else it's their place to carry out some chore or task than the energy it would take to actually complete the task. Families, schools, business, and communities built upon a culture where people focus on doing what is needed regardless of their standing and status are invariably going to be places that are more productive and more pleasant places to work and live.

Prejudice is a learned perspective, not infrequently reinforced through a limited number of personal interactions. Individuals often fail to realize the consequences that bigotry and hatred have on our individual being. A heart that holds onto resentments, prejudices, and hatred is a heart that will be stifled from enjoying the love that leads to happiness, contentment, and enjoying the rich beauty that surrounds us. Fighting to remove these traits and weaknesses is at the heart of

Buddhist teachings and other religious admonitions, for such efforts contributes greatly to our progress spiritually. The effort to continuously improve the emotional condition of our heart and mind not only affects our emotional well-being but also improves our physical health and interpersonal relationships. And that reward is as great as any other pursuit we can strive for.

More Buddhist Teachings on Benevolence and Love

> There is no path to happiness: happiness is the path.
> The price of freedom is simply choosing to be.
> —Buddha

One of the more beautiful aspects of Buddhism I find especially appealing is the emphasis placed on constant self-evaluation and improvement. For those of the Buddhist faith, there are four elements of love that make it true, and followers are encouraged to use meditation as a tool to make progress in embracing the essence of these elements, for in so doing, one's progress on the path of love will be evident. They are not magical concepts; rather, they are simple behaviors and virtues one must hold in order to strengthen love. These four elements of love are not always easy to manifest, but they make relationships much more joyful and fulfilling:

1. *Maitri*—*Maitri* literally means kindness or benevolence. Sometimes an individual may want to love someone but fails to truly understand that individual's deepest needs and desires. *Maitri* involves seeing the person as they are and seeking

to understand them. Only through understanding them, can we understand how to support and encourage them. Part of overcoming prejudice is rooted in making a conscious effort to understand cultures and peoples we differ from, and how they see the world. Kindness impels one to not make rash judgements, but to be patient and kind in dealing with those we may not fully understand.

2. *Karuna*—The second element of true love is *karuna*, meaning compassion. The essence of this principle lies in cultivating the desire to ease the pain of others. This principle is also based on seeking greater understanding of others, especially the suffering they experience. Only when we truly understand others' challenges and pain can we fully know how to work to alleviate it. The practice of meditation is a tool to aid our understanding of those we interact with. Our focus should be on meditating to better empathize with what troubles them and how we can be of greater support. In this way, meditation is a powerful tool to achieve greater understanding of others, as well as the actions we can manifest to assist them.

3. *Mudita*—*Mudita* translates as joy or happiness. If there is no joy in our relationships, there cannot be true love in them. The world is full of individuals who take themselves too seriously. If a relationship upsets us chronically, there is a lack of love in it or the love has been lost. Love is fulfilling and brings joy and happiness to those

who feel it. The joy and happiness we experience from love strengthens our affections. This is a sign the love is real. Regaining joy in relationships that are strained, focusing on the activities that bring those relationships joy and laughter, will invariably lead to improvement in our relationships. Perhaps just loosening up a bit and having a lighter perspective on our feelings of self-importance can ease tension and lighten the stress on our relationships.

4. *Upeksha*—*Upeksha* literally means freedom. Benevolence makes peoples feel free within relationships. We must have the freedom to be ourselves and allow those with whom we interact to be comfortable enough to be themselves. Benevolence does not force individuals to be something they are not. There should also be the freedom within relationships to communicate openly, having the comfort to share ideas and thoughts without fear of judgment. Few things encourage such openness more than sincerely listening. Listening and encouraging open communication is not always easy, especially if an individual is strongly opinionated or has a tendency to complain when communicating. Patience and sustained effort at keeping the lines of communication open are worthy objectives, and the price paid for improved relations is worth our most robust commitment.

The law of karma and the personal benefits of pursuing benevolence

Why pursue benevolence in hopes of overcoming prejudice? Buddhist, Islamic, Jewish, Baha'i, Christian, and other philosophical and religious writings give insightful thoughts on the power of benevolence and generosity to improve the quality of lives of individuals who practice cultivating these qualities. Consider some of the scriptures below.

> A generous person will prosper; whoever refreshes others will be refreshed.
> —Proverbs 11:25

> In everything I did, I showed you that by this kind of hard work we must help the weak, remembering the words the Lord Jesus himself said: "It is more blessed to give than to receive."
> —Acts 20:35

> The righteous care for the needs of their animals, but the kindest acts of the wicked are cruel.
> —Proverbs 12:10

> A heart at peace gives life to the body, but envy rots the bones.
> —Proverbs 14:30

Similarly, the Quran addresses the rewards of generosity:

> Indeed, Allah commands justice, grace, as well as courtesy to close relatives. He forbids indecency, wickedness, and aggression. He instructs you so perhaps you will be mindful.
> —Quran 16:90

> And let not those who [greedily] withhold what Allah has given them of His bounty ever think that it is better for them. Rather, it is worse for them. Their necks will be encircled by what they withheld on the Day of Resurrection. And to Allah belongs the heritage of the heavens and the earth. And Allah, of what you do, is [fully] Aware.
> —Quran 3:180

> Indeed, the men who practice charity and the women who practice charity and [they who] have loaned Allah a goodly loan — it will be multiplied for them, and they will have a noble reward.
> —Quran 57:18

For Buddhists, the law of karma cannot be escaped. For Christians, the principle is similar: we reap what we sow. This principle is present throughout many religious traditions and beliefs in slight variations of word and nuance of idea. But at its heart, the idea of karma, essentially reaping what we sow, is generally a widely accepted belief by large segments of humanity. Its prevalence throughout a diversity of cultures and religious traditions is rooted in the actuality of its value. As physical laws govern the universe and their effects cannot be escaped, so too spiritual laws exist, and their effects cannot be contravened. When an individual cultivates benevolence and practices thoughts and actions in harmony with such, the results will bring happiness and contentment and benevolence to their life in return. This law cannot be escaped, whether for good or for evil. In this way, the practice of benevolence is not one rooted in the conscious, selfish practice of such, but rather a realization that such thoughts and conduct will bring positive results for both ourselves and those we interact with. It is not a justification for our actions but an acknowledgement that pursuing the path of benevolence will ultimately result in improvement of our own life, though such a course may be challenging and require self-sacrifice.

The return of benevolence's fruitage in our lives may not necessarily come from the individuals to whom we've shown such kindness, but the results will invariably blossom in time. As the laws of physics cannot be contravened, this universal spiritual law also must come to fruition. In many ways, the chief benefit rests in the development of proper character and nobility, the wealth of spirit. This wealth will yield fruitage predominantly in our relationships, which are the possessions of greatest value in our lives, but benevolence always leads to tangible beneficial consequences. This is a pattern I've repeatedly seen play out in my own life.

When I was senior in college, I decided to do a clothes drive for the homeless in the communities I'd been photographing. I put large cardboard boxes in the dorms all across campus, and after a couple weeks, collected the boxes and drove downtown to one of the more densely populated homeless areas of Atlanta and distributed the clothes and shoes generously donated. Collecting the clothes was what was within my power to do. I had no other motivation than helping individuals, some of whom had been generous by willingly let me photograph them. I felt content knowing I'd made a small difference in the lives of people I had no obligation to help, sharing the same benevolence I'd personally received from people who had no obligation to assist me.

A few months after I distributed the clothes, I went on my medical school interview. At the end of the day-long interview process, the dean of admissions at the school pulled out a paper and told me I had a decision to make. If I committed right then and there, he guaranteed me a full scholarship to attend the school, but I had to decide that day. I went outside, called my parents to tell them of the offer, and then signed. It was the only medical school interview I completed. While my science GPA was 3.9 and I had good MCAT scores, I didn't feel like I merited the scholarship. There were many applicants with similar credentials, as is the nature of the competitive pool of applicants that comprise any medical school's applicant pool. At the time, I felt immediately like what I'd done collecting the clothes had some bearing on the scholarship being offered.

My intuitions are almost never wrong; it's a gift I've known since childhood. Still, regardless of whether the scholarship being offered was related to my benevolent deed, it demonstrates a pattern of reciprocity, often not coming from the individuals I've directly helped. Such has happened to me more times than I can recount, and this pattern has been repeatedly observed by others in their own lives. In an interview, the famous actor Dany Trejo, commenting on his serendipitous acting break while supporting someone with an addiction problem, recognized this same pattern: "Everything good that ever happened to me in my life was when I was doing something for somebody else."

Prejudice is an indolent poison in the heart. It is often a hidden flaw, concealed from others. It may not always result in the active practice of malevolent deeds against others, but it often does. More often than not, bigotry results in simply not being open to friendships and relationships, or forbidding opportunity or fairness to those deserving. The effort to rid our hearts of this poison is worth every effort, for our emotional and physical well-being are intimately intertwined with the degree of peace and compassion flowing from our heart. The more we embrace benevolence and the impartiality, and the kindness, and love it engenders, the spiritually richer and more contented our life will be. Often we will find ourselves encouraged by the unexpected generosity the universe freely bestows upon those whose hearts align with the glorious energy love resonates, and the benevolence we show, the universe never fails to return multiplied in far greater measure than what our own actions merit.

> Deal ye one with another with the utmost love and harmony, with friendliness and fellowship... This goal excelleth every other goal, and this aspiration is the monarch of all aspirations.
> —Bahá'u'lláh, *Gleanings from the Writings of Bahá'u'lláh*

Chapter Seven
Religion and Humanity's Yearning for Love

Religions are different roads converging to the same point. What does it matter that we take different roads, so long as we reach the same goal?

— Gandhi

This is my simple religion. No need for temples. No need for complicated philosophy. Your own mind, your own heart is the temple. Your philosophy is simple kindness.

—Dalai Lama XIV

Love is my religion.

—Anonymous

I have three lovable, at times very life-sucking, companions society has labeled children. Of course I love them. And anyone who has children or spends any real amount of time with kids knows how vigorously they vie for your attention. "Daddy look at me!" "No Daddy, watch me!" or "Watch what I can do, Dad!" are constant proclamations in my house and in children-filled homes around the world. This phenomenon points to an intuitive truth almost all people acknowledge and yet often still fail to practically understand in a way that improves the quality of their lives and relationships with the people around them: *The need to feel loved and appreciated is the greatest of all human needs.* We see it manifest in childhood, and it is the dominant human drive throughout life.

In children, as in adults, the quest for acknowledgement and value manifests constantly.

Children seek external validation; essentially, they perceive their individual value by how others view them. This is also the view many adults maintain throughout adulthood, albeit with subtle nuances. As children grow to adulthood, they learn they cannot rely on others to satisfy their emotional need for validation based on benevolence and as a consequence develop means of meeting their need for love and fulfillment in other ways.

Love is an addictive feeling. Poems and masterpieces of all sorts, and even wars, have been impelled by it. There is a gross deficiency of love in society. The reality that random benevolence is nearly universally met with skepticism and questioned is a sign of the deplorable state of human society and relations. Most people have great doubt of acts of genuine altruism because of how dominantly self-interest affects humankind. The result is that individuals will gravitate to individuals and activities that satisfy that yearning for love, albeit even if such comes with great destructiveness.

Among the most cherished of all values people hold is their spiritual and religious beliefs, or their choice to have a lack thereof. People have the right to believe and worship freely in most industrialized nations. And while most religions have a version of the golden rule, to treat others as we want to be treated, religions have a tainted history of being divisive—in fact, often violently so.

At its heart, spirituality is an essential pursuit for people. The cultivation of benevolence and kindness is at the heart of inner peace and calmness. It helps relationships and contributes to well-being and even physical health. As a physician often dealing with emergencies and critically ill patients, I've seen countless patients and families cope with the most grievous of tragedies and illnesses. Invariably, the ones that manage those trials and difficulties the best usually have some sort of spiritual inclination and practice. While I don't personally advocate any particular religious sect or spiritual pursuit, I believe that there is good in all religion and that a person can gain happiness and purpose without following a strict system of beliefs ascribed to any particular sect.

I've found kindness and benevolence in all manner of individuals, from every religion and every belief system, including those with the absence of any religious inclinations. In my studies and experience with regard to moral standards and guidance, there is usually very little that separates most religions. The difference between most religions usually lies in beliefs and customs that have very little impact on relationships and on how people should treat each other. This is because at the heart of all religious practice is the sincere commitment to manifesting love. That pursuit of practicing love leads to benevolent deeds, regardless of what religion a person practices.

Often the real problem with religion is many people fail to sincerely apply the teachings encouraged by the tenants of the very religions they claim to practice. It is truly as Gandhi said, "I like your Christ, but I do not like your Christians. Your Christians are so unlike your Christ." And then there is the admonition to not judge others. Among the most basic teachings Jesus taught was to love our neighbors as ourselves. For Christians, using the Bible as their authority, there's no doubt considering others inferior is in direct opposition to what Jesus taught, and also to what his closest followers taught. In his most popular sermon, the Sermon on the Mount, Jesus is quoted as saying,

> [1] "Do not judge, or you too will be judged. [2] For in the same way you judge others, you will be judged, and with the measure you use, it will be measured to you.
>
> [3] "Why do you look at the speck of sawdust in your brother's eye and pay no attention to the plank in your own eye? [4] How can you say to your brother, 'Let me take the speck out of your eye,' when all the time there is a plank in your own eye? [5] You hypocrite, first take the plank out of your own eye, and then you will see clearly to remove the speck from your brother's eye."
>
> —Matthew 7:1–5

And consider this account of Jesus found in Philippians 2:1–4:

> **Imitating Christ's Humility**
>
> [1] Therefore if you have any encouragement from being united with Christ, if any comfort from his love, if any common sharing in the Spirit, if any tenderness and compassion, [2] then make my joy complete by being like-minded, having the same love, being one in spirit and of one mind. [3] Do nothing out of selfish ambition or vain conceit. Rather, in humility value others above yourselves, [4] not looking to your own interests but each of you to the interests of the others.

This account describes Jesus as a person who considered others as superior in value to himself. For Christians, who consider him as the preeminent example, certainly that would be a verse they wouldn't want to ignore. This same pattern of encouraging humility is found in Islam and other major religious texts throughout the world. Consider the Quran. Allah the Exalted said:

وَعِبَادُ الرَّحْمَٰنِ الَّذِينَ يَمْشُونَ عَلَى الْأَرْضِ هَوْنًا وَإِذَا خَاطَبَهُمُ الْجَاهِلُونَ قَالُوا سَلَامًا

[The servants of the Most Merciful are those who walk upon the earth in humility, and when the ignorant address them, they say words of peace.]
—Surah Al-Furqan 25:63

And Allah said:

ادْعُوا رَبَّكُمْ تَضَرُّعًا وَخُفْيَةً إِنَّهُ لَا يُحِبُّ الْمُعْتَدِينَ

[Call upon your Lord with humility and in private. Verily, He does not love transgressors.]
—Surah Al-A'raf 7:55

And Allah said:

وَاذْكُر رَّبَّكَ فِي نَفْسِكَ تَضَرُّعًا وَخِيفَةً وَدُونَ الْجَهْرِ مِنَ الْقَوْلِ بِالْغُدُوِّ وَالْآصَالِ وَلَا تَكُن مِّنَ الْغَافِلِينَ

[Remember your Lord in yourselves with humility and in private without announcing it in the mornings and evenings, and do not be among the heedless.]
—Surah Al-A'raf 7:205

And Allah said:

> وَاخْفِضْ لَهُمَا جَنَاحَ الذُّلِّ مِنَ الرَّحْمَةِ وَقُل رَّبِّ ارْحَمْهُمَا كَمَا رَبَّيَانِي صَغِيرًا
>
> [Lower to your parents the wing of humility out of mercy and say: My Lord, have mercy upon them as they brought me up when I was small.]
>
> —Surah Al-Isra 17:24

> Iyad ibn Himar reported: The Messenger of Allah, peace and blessings be upon him, said, "Verily, Allah has revealed to me that you must be humble towards one another, so that no one oppresses another or boasts to another."
>
> —Ṣaḥīḥ Muslim 2865

This same exhortation for humility is found throughout Judaism and the Old Testament in the Bible. In the Jewish tradition, humility is among the greatest of the virtues, as its opposite, pride, is among the worst of the vices. Moses, the greatest of men, is described as the most humble:

> Now the man Moses was very meek, above all the men that were on the face of the earth.
>
> —Numbers 12:3

The patriarch Abraham protests to God:

> Behold now, I have taken upon me to speak unto the Lord, who am but dust and ashes
>
> —Genesis 18:27

So essential is humility to Judaism and Christianity that Bible writers state that it is impossible to even learn God's ways without such. For example,

> [8] Good and upright is the Lord therefore he instructs sinners in his ways. [9] He guides the humble in what is right and teaches them his way.
> —Psalm 25:8–9

We see the same pattern of emphasis on humility in Buddhism and Taoism.

> The sage puts himself last and becomes the first.
> —*Tao Te Ching*, chapter 7

When commenting on humility in Buddhism, Raymond Lam writes,

> For all his extraordinary achievements throughout his life, Thich Nhat Hanh is extremely humble. For people in Buddhist discipleship, humility liberates them from false perceptions about the world and the intentions of others. It allows them to be more genuine with their fellow students. Being humble in thought, speech, and mind benefits oneself and every relationship one can conceive of, from those with loved ones to mere acquaintances. Therefore, Nhat Hanh is one of the freest human individuals alive because he is, in Martin Luther King Jr.'s words, "humble and devout [my emphasis]". It is no wonder that Thich Nhat Hanh remains an incredibly happy and serene man even after experiencing so much that would break more ordinary individuals.
> —Raymond Lam, "Being Humble Is Itself a Spiritual Practice," *Buddhistdoor Global* (July 25, 2010)

Again and again, in the most prominent religious holy books in the world, we see this recurrent theme of encouraging humility. How could it be that the very beliefs cherished by so many have a history of being used to manipulate and twist individuals into becoming destructive forces among humankind? Is it not because at the core of many religions is the belief in practice that one religion is superior to another? While the Crusades pitted Christians and Muslims against each other for centuries, we still see ongoing tensions in the present day among practitioners of various religious sects.

The threat Islamic extremism poses is real, as are extreme Christian conservative views, the first often leading to suicide bombings and the latter to attacks on abortion clinics and mass shootings at mosques and synagogues. Often, these violent instigators are driven by this same thought of religious superiority, driven solely by belief in the superiority of their religious traditions, and the illusion of superiority manifests its toxicity again.

Buddha never advocated violence. Jesus never advocated violence. Mohammad certainly never advocated the slaughter of innocents. These patterns of religiously motivated violence reveal yet another manifestation of abuse humanity persistently achieves in every aspect of society. Even that which should be viewed as most sacred is not infrequently morphed into an instrument of destruction and violence. I sometimes wonder if religion were solely confined to being practiced within one's home, would the world be a safer place? It is an interesting hypothetical. Interestingly, one Bible prophesy points to a time when all religion will become illegal to practice (Revelation 18). It is a staggering prophecy, largely because it seems so impossible given the nature of religious freedom much of the world now knows.

Whatever comes, the unification of humankind under one global government must mean an eradication of sectarian religious divisions as similarly as our artificial physical divisions too are eroded. At every religion's heart lives the divine mandate for love, and love does not work what is evil to one's neighbor. Perhaps there will be a time when the world is unified in the divine pursuit of love as the penultimate religious pursuit, more than that of any dividing religious customs or beliefs. When that is so and all divisions, both spiritual and physical are removed, then and only then will humanity be poised to reach its optimal potential.

Chapter Eight
Overcoming Tragedy and Disadvantage

You should never view your challenges as a disadvantage. Instead, it's important for you to understand that your experience facing and overcoming adversity is actually one of your biggest advantages.

—Michelle Obama

Although the world is full of suffering, it is also full of the overcoming of it.

—Helen Keller

Helen Keller wasn't born blind and deaf. Her eventual famous handicap developed as a result of a contracted illness, possibly scarlet fever. Regardless of the cause, Keller spent the rest of her life coping with the challenge of having to deal with the world primarily through her sense of touch, living as a blind and deaf woman. It was a disadvantage she overcame gloriously, and she achieved a noteworthy life dedicated to public service. Despite her disadvantage, Keller became a stalwart advocate for the deaf and blind community at a time when women with such conditions were particularly shunned. Indeed, Michelle Obama's words proved true for Keller, and her affliction, and her overcoming of it, led to her becoming the greatest public advocate for blind and deaf persons in history.

While *The Illusion of Superiority* deals primarily with the lack of inherent superiority among individuals, the are no shortages of circumstances in life that place individuals at inherent disadvantage. Handicaps, inherited or acquired; challenges of circumstance, such as parental abandonment or child abuse; the loss of loved ones, either parents, children, or siblings; accidents; and other happenings can scar individuals for life and cause both economic and emotional challenges that can have lifelong ramifications. Even when people are well adjusted to such circumstances, the resentment from such, whether the circumstances come by unexpected change, by error, or by malevolence, often cannot be contravened permanently, and the pain of some things simply cannot stay buried. Constant vigilance is needed to avoid returning to cycles of self-destructive behaviors.

I've known my own share of loss and disappointments, disadvantages brought by circumstance and some by my own naivety and poor choices, but nothing I've ever experienced comes close to the pain felt from losing a child. I feel no need for a "who's suffered more" competition, listing my credentials as a person who's known suffering as a means of establishing my credentials on the matter, as if my losses should somehow be weighed against the losses of others as a way to certify my expertise on the subject. I've found the exercise of even trying to logically explain the *why* of tragedy in any way simply results in a cycle that contributes neither to coping nor to healing, and once the loss is experienced, the only real path forward is to focus all one's energy on building the new and moving forward. Reliving painful circumstances is like driving with one foot on the brake, an exercise that hinders one's path and makes the journey through life needlessly cumbersome and onerous. For any hardship of significant consequence, the aim is to use the pain, to learn from it, and to turn the suffering into fuel to build something of lasting value, something of benefit for society, for oneself and one's family.

Grief's an indolent poison, the cost often no less than the shortening of a person's life through the transitioning to a faster death, a more numb existence, as any number of attempts to quiet the pain of loss can take any number of paths of self-abuse. Promiscuity, drug abuse, alcohol addiction, anhedonia, social withdrawal, and obesity are just a handful of slow deaths the weight of tragic loss often leaves in its wake.

I've known my own periods of dark cycles. For me, Buddhist teachings on suffering, Jesus's words in the Sermon on the Mount, and various meditation practices have been the greatest philosophical perspectives and tool aiding in both healing and well-being in my life. I don't pretend to know what works best for coping, only what's been most effective for me. We'll consider each of these methods, as these are some practical tools I personally know to be effective.

For the Buddhist, suffering is an inherent part of life. That actuality is in itself one of the central tenets of Buddhism. The acceptance of the unavoidability of suffering, to me, is one of the most calming perspectives and philosophical realities a person can embrace of any philosophical or religious tradition. All life is suffering. When this is understood and accepted, an individual ceases trying to make sense of suffering as if it is something that can be entirely avoided, as if something strange and unusual were afflicting them. It is an inherent component of life, as inherent as birth and death are, and it's an actuality all must endure. And while it is true, some suffering is more difficult than others to endure, and some circumstances are more grievous than others. I have found, as have others, that in every circumstance, both positives and negatives exist in equal measure and the counterbalance of harmony remains, as the universal law remains true for every person, in every circumstance.

I've known the joys of having exceptional natural ability and talent in a variety of ways, as well as the suffering of some of the most difficult circumstances and losses one can endure in life. In everything there is balance, both good and bad, in all circumstances. This is also a tenet of the Yoruba faith, to which my genetic heritage binds me; of Buddhism, which I see as closest to my personal philosophy and which binds me; and in some respects, even Christianity, to which my childhood religious instruction binds me. For the Yoruba, for example, good and bad often come in equal amounts; for example, when something good comes, something bad accompanies it. In everything there is duality, and while some circumstances can be alleviated or adjusted with the support of one's ancestors, for example, there is no way to completely avoid suffering. One of the most beautiful tenets in Yoruba is that good character overcomes and makes all curses or magic ineffective. That is, if a person maintains a moral and upright character, no evil directed toward them will be effective. In this sense, as in Buddhism, an individual is motivated to take an active role in their life and accept complete responsibility for their behavior and choices.

In this sense, at the root of every manner of acceptance of tragedy and difficulty, understanding our personal role in moving forward lies at the heart of coping. Even for Christians and others adherent to Abrahamic religious traditions, those who believe their source of strength originates from outside of themselves, there is also the impetus for personal responsibility, principally through prayer and obeying scriptural principles directed in sacred texts.

Throw your burdens on the Lord. With evil things God does not try anyone, in all things by prayer and supplication make your petitions known to God, and the peace of mind that excels all thought will guard your mental powers. Trust in the Lord with all your heart.

Scriptures like these highlight the admonition for adherents of Abrahamic religious traditions to have an active role in making choices that they have personal responsibility and control over, with faith that their God will provide the support and encouragement for them to endure tragedy and/or difficult circumstances.

For the Buddhist, meditation as a tool often lies at the heart of healing and growth. In this regard, the four tenets become essential aims and food for thought during meditation. These are neither esoteric nor empty mediation practices, nor a clearing of one's mind, as it were, but rather a focused, guided mediation, where the emphasis on personal growth and responding in healthy ways with loving kindness and understanding for oneself and others is the ultimate goal. Meditation becomes the daily tool for Buddhists used to restrain themselves from paths that ultimately lead to further cycles of suffering and imbalance.

Learning to control one's breath, emotions, and intentions through meditation is one of the healthiest ways of coping with loss and tragedy. No one can erase the pain of loss for us. It is a burden we must learn to bear through the practice of habits, philosophies, and practices we find most supportive.

While meditation is often thought of a practice predominantly associated with Eastern religious practices, meditation is also a practice heavily encouraged for those of Abrahamic religious traditions in the Bible.

> [11] I will remember the deeds of the LORD; yes, I will remember your miracles of long ago. [12] I will consider all your works and meditate on all your mighty deeds.
> —Psalms 77:11–12

> But whose delight is in the law of the Lord, and who meditates on his law day and night.
> —Psalms 1:2

> I meditate on your precepts and consider your ways.
> —Psalms 119:15

> Within your temple, O God, we meditate on your unfailing love.
> —Psalms 48:9

> Though rulers sit together and slander me, your servant will meditate on your decrees.
> —Psalms 119:23

> Oh, how I love your law! I meditate on it all day long.
> —Psalms 119:97

> I have more insight than all my teachers, for I meditate on your statutes.
> —Psalms 119:99

> He went out to the field one evening to meditate, and as he looked up, he saw camels approaching.
> —Genesis 24:63

> Cause me to understand the way of your precepts, that I may meditate on your wonderful deeds.
> —Psalms 119:27

> I reach out for your commands, which I love, that I may meditate on your decrees.
> —Psalms 119:48

> May the arrogant be put to shame for wronging me without cause; but I will meditate on your precepts.
> —Psalms 119:78

> My eyes stay open through the watches of the night, that I may meditate on your promises.
> —Psalms 119:148

> I remember the days of long ago; I meditate on all your works and consider what your hands have done.
> —Psalms 143:5

> They speak of the glorious splendor of your majesty—and I will meditate on your wonderful works.
> —Psalms 145:5

> Keep this Book of the Law always on your lips; meditate on it day and night, so that you may be careful to do everything written in it. Then you will be prosperous and successful.
> —Joshua 1:8

The Quran also mentions the power of meditation to yield benevolent fruitage, though the Islamic faith is seldom recognized and credited for encouraging meditation. Consider these verses from the Quran:

They are steadfast, truthful, submitting, charitable, and meditators at dawn.

—Quran 3:17

They are the repenters, the worshipers, the praisers, the meditators, the bowing and prostrating, the advocators of righteousness and forbidders of evil, and the keepers of GOD's laws. Give good news to such believers.

—Quran 9:112

During the night, you shall meditate for extra credit, that your Lord may raise you to an honorable rank. And say, "My Lord, admit me an honorable admittance, and let me depart an honorable departure, and grant me from You a powerful support."

—Quran 17:79-80

The worshipers of the Most Gracious are those who tread the earth gently, and when the ignorant speak to them, they only utter peace. In the privacy of the night, they meditate on their Lord, and fall prostrate. And they say, "Our Lord, spare us the agony of Hell; its retribution is horrendous. It is the worst abode; the worst destiny."

—Quran 25:63-66

And put your trust in the Almighty, Most Merciful. Who sees you when you meditate during the night. And your frequent prostrations. He is the Hearer, the Omniscient.

—Quran 26:217-220

Is it not better to be one of those who meditate in the night, prostrating and staying up, being aware of the Hereafter, and seeking the mercy of their Lord? Say, "Are those who know equal to those who do not know?" Only those who possess intelligence will take heed.

—Quran 39:9

> ... praise and glorify your Lord before sunrise, and before sunset. During the night you shall meditate on His name, and after prostrating.
>
> —Quran 50:39–40

While the emphasis on nonreligious forms of mediation is often on the individual and how one responds to a variety of challenges and difficulties, for those of Abrahamic religious traditions, the emphasis rests on remembering God and his activity. Often there is impetus to remember God's actions on behalf of his worshippers in past times, as well as on remembering his laws and commands. In so remembering and cogitating, such practice is intended to strengthen one's faith and obedience, which are often severely tested when experiencing challenges and difficulties. By remaining obedient to the commands issued in God's word, for adherents of Abrahamic religious traditions, that obedience will prevent the adoption of habits and practices detrimental to their health and welfare, such as anger, violence, promiscuity, drug abuse, and alcoholism, for example. James 1:13–15 encourages Christians to remember that God himself is not the source of one's problems and difficulties:

> [13] When tempted, no one should say, "God is tempting me." For God cannot be tempted by evil, nor does he tempt anyone; [14] but each person is tempted when they are dragged away by their own evil desire and enticed. [15] Then, after desire has conceived, it gives birth to sin; and sin, when it is full-grown, gives birth to death.
>
> —James 1:13–15

For the nonreligious person, such scriptures have no significance and meaning. Still, the practice of meditation as means of coping has great significance, as does the adoption of habits that encourage health, well-being, and the avoidance of social isolation.

    Clearly, every person must find for themselves those mechanisms that allow for coping with loss and tragedy in healthy and balanced ways. Life goes on, as do our responsibilities and the need to care for ourselves and family and meet our financial and other obligations. Whatever course we pursue, our focus on activities that alter our attention to more selfless priorities ultimately makes the burden of moving on easier to bear. In so doing, we honor all life and the circle of those who know our affection, and in so doing, add meaning and purpose, elevating our life from mundane patterns of merely existing.

Chapter Nine
Intuition vs. Artificial Intelligence

The opposite of a correct statement is a false statement. But the opposite of a profound truth may well be another profound truth.

— Niels Bohr

The core practice of magic is: The execution of a willed intent to create change in the material world, which either defies, hastens or purifies the consequences of natural cause and effect.

— Zeena Schreck

Isaac Asimov is my boy! When I first saw the trailer for Apple TV's new series based on his *Foundation* novels, I crammed finishing the first three of the seven novels in about ten days, before disappointedly learning the show would be only loosely based on the books. I'd always been a fan of Asimov, though not terribly familiar with any of his books outside the Robot series.

iRobot, based on Asimov's Robot series, is still on my list of favorite movies and books. A robot army lead by a malevolent artificial intelligence (AI), well you had me at hello! Science fiction, when done well and rooted in reasonable theoretical possibility, is one of the most intriguing genres in existence. And Asimov, the official grandfather of the genre, certainly knows his way around gripping concepts and characters.

I was a just sophomore in college when I first read another brilliant intellectual giant's words, Nobel Prize–winner Bohr's quote above. I found the quote in the textbook for my general chemistry class, but his words immediately resonated as having considerably more import beyond the realm of science and chemistry. The statement instantly struck me as preeminently profound, and I never forgot it. It would be more than two decades after I first read the quote before I fully understood what its ramifications meant for my work personally, specifically on my thinking regarding technological advancement, society, and the existence of unseen entities. Principally, it came to guide my thoughts on what understanding the full intuitive capability of the human brain means as a form of technological innovation, and how sustainable technological integration must be achieved if our species is to survive and biodiversity as we know it is to continue.

    With all that humanity has achieved technologically, our species has failed flagrantly in our most essential task. As a species, we have an obligation to be moral stewards of Earth and its resources. Humans can affect change to our environment in a way no other organic life on our planet has the capacity to, but we are not above other life-forms in our dependency upon Earth for our survival. This position impels us to live in harmony with the natural world, which sustains not only our existence but that of every species of life on our planet, much of which has direct use in providing for our health and well-being. Many plant and animal species offer useful products to humanity, and we've failed to discover the full ramifications of much of the potential present within the natural world. What medications, processes that could provide useful biomimetic patterns, fuels, industrial products, and other beneficial derivatives might exist within the rainforests, coral reefs, volcanoes, sea floor, Arctic, and other areas that have not fully been understood? Many of the shamans I know believe that within the natural world exists the cure to every ailment humanity experiences. Perhaps they are right. But I fear we might never know whether such is true because of the pattern of careless stewardship our species has persisted on for the past several centuries.

Around Earth, the extinction of enormous varieties of species is occurring at an alarming rate. Some calculations place the current rate of extinction of various plant and animal species at twenty to two hundred extinctions per million species per year, a significant increase from the approximate one extinction per million species per year believed to be the rate when discounting the effect of human activity. Due to economic pressures for nonessential products like palm oil, the rainforests in large parts of the Amazon, and other parts of the world, are burning at a rate of thousands of acres a day. Plastic bottles and other nonbiodegradable products fill landfills, oceans, and rivers. Our climate, the delicate dance of various cycles of precipitation, wind, and temperature, grows more unstable as ice caps melt and fossils fuels poison the air and ground. Humanity, stuck in the cycle of consumerism as recreation, is mostly apathetic and uninspired to meet the challenges, or ill equipped to do so. The significant societal changes needed to quell the alarming rate of decline of the ecosystems are left unfulfilled as humanity goes about the routine of consumerism and exploitation of the very environment that sustains our existence.

We live upon the precipice of one of the most significant technological accomplishments in the history of human civilization. The development of sentient AI, AI that has self-awareness and the ability to make moral judgements, will likely occur during my lifetime. Its integration into society will profoundly affect my children's lives as it becomes a part of virtually every commercial endeavor within civilization. I find it useful to think of AI as essentially the production of a digital life-form by humanity, and it must be thoughtfully considered from the seventh-generation principle, namely, what will AI become in 140 to 200 years? In many respects, it's a technology that will take on a life of its own by its very nature. And as such, it's impossible to predict how its life will evolve and grow. Regardless of how well-intentioned AI's production is, no computer scientist can fully understand what the ramifications of AI's evolution will mean for humanity and our environment in seven generations. AI's development of self-awareness is glaringly problematic, for if that self-awareness means comprehension of the magnitude to which humanity's activities threaten not only our species' existence but the existence of every living thing on this planet including the AI entity itself, what might the possible response of a sentient AI be?

AI represents the pinnacle of humanity's external technological advancement. I say external because it's an advancement based on materials external to the human body. Thinking about sentient AI from the perspective of the seventh-generation principle, it's obviously a technology whose maturation and proliferation will in time be entirely independent upon humanity for its sustenance. The integration of 3D-printer technology, solar-power technology, and robotized mass-production manufacturing, represents the potential for AI self-replicating and making other modifications entirely without human input. The potential dangers of such to the human race are obvious, especially should a powerful, sentient AI consider humanity a threat to its existence. The integration of AI into military infrastructure, weapons manufacturing, the financial system, food production and distribution, water treatment and distribution, and other arenas will give AI the potential to exert immense control over large sectors of the populace. How these systems are regulated and integrated and what safeguards are put into place to protect humanity's welfare is obviously essential and is just as important as the development of sentient AI itself. It's not been the pattern of human technological advancement to give significant enough forethought to the ramifications of how such technology will affect society and the environment. I gave considerable thought to this as I considered sharing the knowledge of unseen entities' existence and what the things individuals can accomplish with them means for society.

There are a great many grave concerns surrounding the generation of sentient digital life-forms. Will sentient AI manifest the same pattern organic life displays to protect its survival? Will it follow the pattern of humanity and seek its own liberty and freedom? What might such a desire for liberty mean for human civilization? Will AI have a desire to reproduce itself?

What will the moral compass and ethical fabric of sentient AI be? What dangers might superintelligent, amoral digital beings integrated in significant ways with vital infrastructures hold for the world? Will sentient AI consider the human race a threat to its survival? These are not questions that can be answered easily. But clearly, even a brief consideration of them raises concerns that are beyond vital for humanity to consider, aside from the apparent loss of jobs increased robotization will mean for society. The very balance of human existence could be at stake.

In as much as AI is poised to bring drastic changes to virtually every endeavor in human society, so too the knowledge of unseen entities and their enormous capabilities represents similar potential. As AI represents the pinnacle of external human technology, shamans and others who interact with unseen entities represent the pinnacle of internal human technological advancement. The ability to connect symbiotically with these entities, and use such interaction for an individual or community's welfare, represents enormous potential for human society as a whole. Still, like AI, it is a technology that is not without substantial potential dangers.

In *The World Hidden*, the eyewitness accounts I collected and shared focused primarily on what some would characterize as *white magic*. Terms like *magic*, *psychic*, *shaman*, *witch*, and the like are antiquated words that describe individuals' ability to interact with unseen entities in various ways. I prefer instead to refer to such activity as the technology of the Indigenous, or TOTI. It's been the pattern of virtually every Indigenous culture that has ever existed to both understand these beings' existence and have the ability to interact with them. This awareness has not entirely benefited humanity. A considerable segment within the community of individuals who possess the knowledge of unseen entities use such knowledge primarily for selfish reasons and have no compunction injuring or killing others as their desires dictate. There is no reason to conclude that others among humanity might not very well imitate such a course, knowing no restraint in applying the powerful abilities these entities possess for unscrupulous intentions. Among the entities themselves, there appear to be both predominantly malevolent and benevolent entities, and considering the enormous capacity of such entities, the dangers they present to humanity are enormous. So why share such knowledge? Why expose this incredibly powerful technology when humankind has shown no restraint for a predilection to every manner of egregious atrocities?

TOTI developed from Indigenous peoples. It was rooted in benevolence for the predominant number of these cultures, and focused on the integration and interdependence of humanity and unseen entities with the natural world. In ways that are still scientifically unclear to me, the natural world provides strength and sustenance to these entities. There was an emphasis on harmony with the natural world, a harmony that contributed to equity not only between humanity and these entities but also between all human members within these Indigenous communities. There was an awareness of the need to preserve the natural world.

These cultures' way of life set a pattern worthy of imitation in living sustainably in harmony with the environment, in no small part because of their understanding of the existence of these entities and their influence. In stark contrast, the exploitative nature of external technological advancement has been an unrelenting pattern of development with little consideration to the impact such technological advancement will have on society or the environment.

Consumerism fueled by technological advancement is a plague upon our planet. And there's no sign humanity will make any drastic change away from the addiction to buying things needed to produce a way of life sustainable for the indefinite survival of not only our species but every other species as well. Extinction is pending for large varieties of Earth's species, and it seems only catastrophe is likely to bring balance to a rapidly decaying system. I believe such change will happen not by choice but by force as plagues and natural disasters continue to escalate and exert pressure on humanity. A significant epidemic, one that kills a large portion of humanity, is not beyond the realm of possibility, as COVID-19's impact disturbingly illustrates humanity's fragility in the face of infectious disease.

Imagine the impact COVID-19 would've had if its mortality was 15 percent or more, instead of the approximately 1 percent worldwide. COVID-19 will not be the last pandemic the world sees. There are a considerable number of shamans who believe that COVID-19, and other emerging plagues, including increasingly drug resistant established diseases, is the response of a planet under siege from humanity. Such thinking intimates that Earth has a form of consciousness. There's no evidence of that so far as science is concerned, but it's an interesting concept, nonetheless. When COVID-19 emerged, global CO_2 emissions plummeted as international transportation activities froze from the shutdowns that occurred around the globe. In a real way, a scientifically measurable way, the effects of reduced human activity led to a drastic short-term change in the poisoning of our air. For some, it was a warning, a plea from Earth for humanity to change its course.

When I consider AI, its enormous capacity for both benefit and harm, I fear its dominant application will be one directed for corporate interest, not humanity's, and greed will simply lead to AI optimizing an already harmful system rooted in consumerism, only more effectively aiding its further exploitation of Earth's resources. A change to the system, something powerful and stark, is needed to shift humanity's mind-set from the mundane existence of merely acquiring more and more goods to activities more in harmony with the genuine lasting benefit of humankind and the natural world.

Exploration and study of TOTI offers humanity a chance to embrace something far more significant. Indigenous technology, rooted in understanding the delicate balance between humankind, unseen entities, and the environment, might provide the shift society needs to embrace a more sustainable way of life. Aside from the powerful abilities unseen entities manifest and the breadth of diversity practical applications of TOTI have for humanity, the real quest humanity will be forced to embrace as a result of understanding unseen entities is about the nature of these beings' origin and the universe's itself.

These beings have an origin. That origin, and the degree to which humanity can understand these entities, will have enormous ramifications for science, religion, philosophy, and every other pursuit of humankind. I cannot ignore the connection these beings have with various occult and religious texts' descriptions. Elements of what is commonly known as ceremonial magic involve the use of religiously infused invocations of entities. Descriptions of unseen entities in religious holy books date back thousands of years before their acknowledgment by Western science, and understanding unseen entities takes on a higher connection with theology in a way that traditional thoughts on alien life aside from such do not.

In the same way AI will evolve in time to take a form that is difficult to predict fully at present, TOTI has evolved throughout generations, far from its original use by Indigenous peoples. In ceremonial magic, the ability to interact with unseen entities became interwoven with religious ideas. Ceremonial magic, what the Indigenous often saw as a path for a synergistic relationship with these entities, became a tool for controlling malevolent entities and impelling them to achieve whatever ends they desired. The connection between honoring nature and living in harmony with these beings was stripped away, and rituals became nearly entirely focused on achieving the desired end. From an objective perspective, such amounts to a Westernizing of TOTI, at least in practice, with the emphasis less on community and harmony and more on exploitation and manipulation. In this way, it's understandable why some have developed an antipathy to a community that is often misunderstood. The exploitation of any technology for selfish reasons ultimately leads to corrupt results for society and individuals personally. This is not to say ceremonial magic is evil in itself. Still, like any powerful technology, individuals within any given community often manifest conduct that can reflect poorly on that community as a whole, without that community being worthy of condemnation.

Sentient AI will be developed. That is not science fiction. Viewed from the perspective of the seventh-generation principle, what human can definitively predict what AI will become in 140 to 200 years? No one. Its development is inevitable, and that actuality and eventuality necessitates that every effort is made to place as many safeguards as possible to limit its potential for destructive and malevolent abuses and uses. The development of strategies to increase the likelihood that the sentient digital life-from that develops will have a code of ethics and morality, one that esteems humanity and the environment, needs to be prioritized as much as the development of the technology itself. It is a technology that represents the apex of human technological development and that holds enormous promise and risks.

In many ways, the study of unseen entities represents something similar for humanity. Such research is already being done and has been done for decades, though in secret. The widespread study of unseen entities poses enormous risks, especially if individuals experiment independently in ways that expose themselves and society to the dangers these entities present. Organized, thoughtful, well-planned research is the only ethical course for studying in any area of technological exploration that involves serious risk. The study of unseen entities is no different. In *The World Hidden*, I spoke of an international multidisciplinary institute dedicated to the research and understanding of unseen entities and the ethical application of TOTI to the benefit humanity. I am convinced that unseen entities hold the key to understanding the universe in ways that even AI cannot not achieve. The added caveat of TOTI's relation to theology is not a concern that should delegitimize the value of its study, nor should it cause one to see such research as lacking objectivity. It should be embraced, for understanding the universe, and whatever beings and forces exist, is of unquantifiable benefit, whatever such study reveals.

    When I finished *The World Hidden*, I sent copies of the manuscript to some of my most trusted family friends, among them a close family friend who is a sincerely religious, elderly physician and another nonreligious friend who is a lawyer. I wanted the opinion of people from various religious perspectives and scientific education levels. I knew the material would be difficult to digest for some of the general public, especially those with strongly entrenched religious views. It was precisely for that reason I wanted their perspectives. I didn't run from the difficulty the material might hold for them; I needed to embrace that difficulty. Only in embracing conflict can a path to resolution, or at least an understanding, be achieved. They both had responses that I hadn't anticipated. The religious physician immediately saw the logic of my argument and the book's value, both scientifically and medically. The nonreligious lawyer told me the book made her want to read the Bible more, as she was unfamiliar with the many scriptures that referenced unseen entities in the Bible. They both had the exact opposite view of what I expected, and it showed me something I hadn't considered when I was writing the book.

  The study of unseen entities, their existence, and their abilities relates to what is described frequently as angels and demons in both the Bible and occult reference texts. Could unseen entities, whose powers I've observed and experienced firsthand, be the angels and/or demons described in these texts? On the surface, scientists and rational thinkers often reflexively dismiss the concept of God, but it may be true. How can anyone declare with certainty the nature of the universe's origin when the vastness of all that is still unknown is beyond humanity's ability to quantify? The accuracy of descriptions of unseen entities within occult texts and the Bible gives these texts an authenticity and credibility that cannot be denied in light of legitimate observations of these entities' abilities. If these entities, often labeled as angels and demons, are simply misunderstood alien life-forms and research reveals that their origin is entirely unrelated to some divine creator, that too would be invaluable. Religious and occult texts could merely be explaining these beings in the realms of a divine because they had no concept of alien life. Regardless of the actuality of their origin, understanding such should at least be attempted. The ramifications for society would be incalculable either way, whatever such study reveals.

Billions of dollars are spent annually on the exploration of space and the development of AI. For a mere fraction of that expense, answers to the nature of humanity's, and the universe's origins could be achieved, and perhaps achieved relatively quickly. There is a community of individuals who already understand both how to communicate with these entities and how to cooperate with them for a diverse array of activities. Understanding how these individuals can see unseen entities, and channel them and their abilities, and what biometric changes occur during such activities has application in various human endeavors. It may help uncover elements of the physical universe and the human mind and body's potential in ways that we simply cannot comprehend at present. The understanding of energy from studying sentient beings without physical form could provide an avenue for achieving a renewable source of clean energy.

In proving these entities' existence, humanity has taken an enormous step in understanding God's existence or nonexistence. The ramifications of that may be galvanizing for human society and may alter our species' self-destructive path. We are not alone in the universe as intelligent sentient beings; in fact, we are significantly inferior in numerous ways to these beings. Their knowledge of the universe represents a technology of far greater consequence than AI ever could offer, for they have an awareness of the universe that possibly dates back billions of years if various texts are accurate in their description of them. The value of understanding them, and responsibly pursuing such a course, must be an endeavor humanity seeks for its own welfare. The future survival of our species, our planet, and the biodiversity that comprises it may significantly depend on recognizing the importance of the unseen entities that surround us.

Chapter Ten
Race and the United States

Indeed it has been said that democracy is the worst form of Government except for all those other forms that have been tried from time to time.

— Winston Churchill

You and I have never known democracy — all we've known is hypocrisy.

— Malcolm X

The oppressed are allowed once every few years to decide which particular representatives of the oppressing class are to represent and repress them.

— Karl Marx

A battle for the soul of White children — that's what the media-driven much-ado-about-nothing burger Critical Race Theory has become in the minds of much of the American consciousness. The recent flurry of media coverage surrounding the theory has completely failed to highlight the theory's primary application has nothing to do at all with how young children are educated. Now ask yourself, why is the media making such a glaring oversight? and perhaps, what does such say about the power of political and corporate interests to influence media content?

Still, though rooted in artifice, the media coverage of the theory is engendering sincere questions about how we educate children and molding public opinion in that regard, resulting in legislative consequences that have real world impact, at least on the local and state level. In Tennessee recently, a law passed that barred teachers from teaching anything in history that would make a student "feel discomfort, guilt, or anguish." I mean, really?! There are no end of factual historical actualities that appropriately should make a person feel discomfort and anguish, for history is really nothing more than the study of the cycle of struggle between one class or people against another. What is our society to become if we are so sensitive to people's feelings that we cannot teach our children the truth? Critical Race Theory has become primarily an argument about how should we color the way we teach history, when the real question we should be asking is how can we teach history more effectively. History speaks for itself, and what we should be considering is how can we more effectively use modern technology to teach it.

    Should kids learn about American History? Yes. Should slavery, the Native American Genocide, Jim Crow and Civil Rights be included in American History? Yes. Should these and other ugly realities of America's history be taught honestly, as well as how their consequences affected Black Americans? Yes, absolutely. To not do so is dishonest.

    America's history needs to be taught, and taught accurately, for both America's successes and struggles provide vital lessons for all. American history is no condemnation of the White race, as if White Americans are some how more morally repugnant than other races, as some would claim them to be, and as some would articulate the lens of Critical Race Theory paints America as. Yes, slavery was a morally repugnant institution propagated overwhelmingly by White Americans, but, so too, the Abolitionist movement to end slavery was also overwhelming lead by White Americans.

When we look at the Civil War, the number of White Union Soldiers Killed in action: 100,000, total White Union Soldiers dead: 320,000 Wounded: 275,200, and nearly 40,000 Black soldiers died over the course of the war—30,000 of infection or disease. The overwhelming number of Americans who fought and died to end slavery were White. The Underground Railroad, the Civil Rights Movement, and even Black Lives Matter, were and have been successful in large part because of the efforts of White America's support of them. So is the pattern of an honest assessment of history; both the successes and struggles of our ancestors are instructional. An honest study of American History doesn't condemn the White Race, or America as some inherently racist country as some fear, rather, balanced study shows both the good and bad of Americans of all ethnic diversities. We cannot pick and choose what to tell of history, nor should we color it to fit the narratives we wish to paint.

We tell the truth, and in so doing, the virtues and failures of our species become readily apparent, for there is both good and bad in great measure in all of humankind. If we avoid the dirty realities of the past, we will miss the great noble deeds that often served to counter balance such, deeds which reveal the true nature of humanity; there is both good and evil in all.

Educational systems are largely controlled by local politics. Local school boards set educational policy. While the federal government provides broad oversight, what is taught at local levels in classrooms is predominantly left to the auspices of local education officials, and as such, the hoopla surrounding critical race theory has little practical effect on the way children are educated throughout the whole of society, nor is it a theory ever intended to do so.

Red states will be red, and blue states will be blue. So, in practice, states historically red or blue will have school boards largely mirroring the politics of each individual locale. While all the media coverage is interesting and all, in effect, it will alter society in the way social media caters to already established preferences, for no federal mandate surrounding Critical Race Theory will ever take place in practice because of the controversy surrounding it. The fact that media outlets are intimating Critical Race Theory has anything at all to do with the education of school-age children reveals a gross misunderstanding of the theory's origin and purpose.

The media is now predominantly a for-profit industry. Let that sink in. And the media, as a for-profit industry, is no different from any other business. While there is the guise of providing unbiased coverage of current events, the media has as much a vested interest in driving viewers to its stations and webpages as any other business. As such, media outlets have an enormous incentive in generating controversy, for controversy is as great a fodder for attracting viewers as anything else. Viewed from the actuality of what the media has become in modern times, a consideration of Critical Race Theory in practice becomes far less caustic.

President Biden is a moderate Democrat, and as such, his policies, to a large degree, aren't going to elicit a visceral condemnation from more reasonable Republicans and independents. That actuality is the reason Right-leaning news media outlets have instead focused more on highlighting his perceived age-related decline than criticizing his policies. If you are a Republican and you have a vested interest in keeping your voter base angry and engaged in an election cycle in which Donald Trump isn't on the ballot, you can't do that from rage driven by Biden's actions, which have largely been more about keeping the ship of America's economy afloat and managing the COVID-19 pandemic, practical matters any reasonable voter won't have tremendous trepidation about. What you need is something far more incendiary, something that strikes at the heart of Trump voters' motivation, hence the manufacturing of the media-driven controversy surrounding Critical Race Theory.

At its heart, the controversy surrounding critical race theory is meant to drive Republican anger and fuel media cycles. It's just another play on race in actuality. And the cycle of manufactured controversy continues as Left-leaning media outlets benefit too from the ongoing conversation about what is at its heart: a very reasonable and practical theory.

Critical Race Theory is a theory that was developed for law students. It spoke nothing about how children should be educated. Its intended focus is on how laws often have unintended consequences and ramifications for minority populations in their practical application. That is in itself a very obvious actuality when any impartial analysis of American history is made. As such, the theory intimates nothing about how we educate young children, or even high schoolers, or even college students, for that matter. It made no intimation that America is an inherently racist country or that such should be taught to children. The aspect of media coverage pushing such thoughts surrounding Critical Race Theory has been completely fabricated, with the sole intent of fueling media cycles and manufacturing anger to keep Republican voters angry and engaged.

Is America inherently racist? It's neither a helpful nor practical notion to categorize America as such, as to do so intimates that Americans by and large are racist, which I personally do not believe to be true. Another added caveat to that is most Americans aren't in a position to alter the lives of others in any significant way even if many are racist, other than through individual acts of violence. Is there prejudice in America? Yes, and prejudice is manifested by some individuals from every race of people in this country. Because Caucasians occupy the greater number of influential positions in society, prejudice by them is going to affect individuals and society to a more consequential degree on average, but persons from every race manifest prejudice.

Are there unparalleled opportunities in America for every race of people? Yes, that is also true. Cream rises to the top. And while opportunity is not equitable in America for a number of reasons—such as consequences from generational cycles of poverty, unequal educational resources and access to quality education, inequities in health care access, racial profiling and unequal sentencing in criminal cases, reduced access to capital for minority small business owners, and other factors surrounding social mobility—a conquering spirit can overcome disadvantage. For while society may display partiality, nature does not, in so far as talent and ability are concerned, and talent, drive, and ability are present in profound measure in every race comprising society. This is not to say victims of oppression can overcome any disadvantage through hard work alone, that society should not make every effort to move in a more egalitarian direction, or that credit is inherently due those positioned above others socially and economically within society, many of whom derive their advantage solely by the inheritance of generational wealth they had no role in building. Rather, through a multitude of avenues, the universe often provides a path forward for those persistent enough to explore various paths.

> We hold these truths to be self-evident, that all men are created equal, that they are endowed by their Creator with certain unalienable Rights, that among these are Life, Liberty and the pursuit of Happiness. The pursuit of our basic needs requires the existence of laws and the physical means for their enforcement. The existence of a significant portion of humanity who lack empathy and have no concern for the welfare of others necessitates a need for the establishment of a framework to protect those who lack the ability to protect themselves. Individuals who possess no significant moral guide to prevent them abusing others and with no regard whatsoever for the well-being of their fellow people exist throughout all socioeconomic and educational levels of humanity. There is an ever-

present awareness in the consciousness of humanity of the threat of the evil of individuals among us.
—Preamble to Declaration of Independence.

The Declaration of Independence is one of the quintessential literary and political documents ever written. It is a marvel for its beauty and moral fortitude, and yet without question, it is a document full of irony at best, or outright hypocrisy at worst. Three of the five men who penned its words, and forty-one of the fifty-six who later signed it, were documented slave owners. Whether you see this reality as irony or hypocrisy, or something in between, the United States' promise is one rooted in ideals its very founders, by their way of life, failed to uphold.

The founding fathers of the United States were as flawed as all humankind. Within their noble and good intentions, their designed new government's architecture provided for the security of their inherent self-interest in protecting and building their own wealth. The foundation of the US economy was established with the brutal institution of slavery. That is incontrovertible. The deliberate barter and trafficking of humans considered chattel provided the backbone of a fledgling nation's material prosperity and economic stability. The consequent periods following slavery—Jim Crow (legalized segregation), the KKK, the 1960s civil rights movement, the treatment of Blacks in the Vietnam War, voter suppression, the modern-day Black Lives Matter movement for social change and criminal justice reform, police brutality, the modern-day White supremacy movement, and racial economic inequality—all reflect America's historically persistent and ongoing racial problem. The callous and cruel murder of George Floyd seemed an eerie version of a public lynching in modern times. It was just one of the countless examples of violence by an agent of the government on a person of non-White descent.

From its inception, America has failed the very creed of its founding principles. The illusion of a harmonizing melting pot of people, where all races are welcomed with equal opportunity, is a fallacy of American idealism that has never had any root in reality for minorities in this county. While European White immigrants were accepted with open arms in America, that has never been true for people of non-White descent. From its founding to the modern day, minorities have never known equal footing in the American Dream.

> It is not your environment, it is you—the quality of your mind, the integrity of your soul and the determination of your will—that will decide your future and shape your life.
> —Dr. Benjamin Mays

Still, even with its history of bigotry, frank discrimination, and disenfranchisement of minorities, America is a land of unparalleled opportunity for all people. Despite inherent disadvantages, a better future is possible for minorities and immigrants in America. Its history of prejudice and discrimination has not prevented many with significant disadvantages from achieving their dreams of a better future for themselves and their family.

Hard work, persistence, determination, and natural talent are the great equalizers, and a conquering spirit here can know unbounded success, despite any disadvantage. And race and poverty, in themselves, are not excuses preventing greatness and prosperity. This is the true jewel of American democracy.

The US Constitution is a marvel of beauty for our nation, for within its thoughtful construction rests the mechanisms by which the aggrieved within society have a voice to push for change, for a fairer system, a more perfect union. Slavery ended by law. Segregation ended by law. Voter rights improved through changes in the law. These advancements were not made without the blood and sacrifices of many, including to a significant and vital degree support from White Americans, despite considerable pushback from opposing institutions and individuals deliberately fighting against the flagrant inhumane treatment of their fellow citizens. *Brown v. Board of Education*, *Plessy v. Ferguson*, and *Citizens United v. FEC* all reflect potent forces of opposition to what is obviously in the egalitarian spirit of fairness. Repeatedly, US history reveals the need for citizens to obtain legal precedence through the US justice system for what the dictates of conscience and basic human principles of fairness and decency should intrinsically dictate.

This actuality illustrates the essence of why moral government is needed. Government's true purpose is to protect its citizenry from itself. The intentional usurping of others' rights is a regularly recurring theme throughout humanity's history. The government should be the ever-present bulwark protecting its citizenry from the masters of mankind's cruelty. Yet it has always been a tool for accomplishing the purposes and aims of a small sector of an opulent few in society. This pattern has revealed itself in every form of government that has ever existed in the history of the world.

From its very inception, American Democracy had the masters of mankind's interests in mind, namely protecting the material prosperity of the wealthy elite, the aristocratic class known as America's founding fathers. Both James Madison and Aristotle identified the same significant problem with democracy. In theory, the core of democracy is expressing the majority's will, which creates several inherent weaknesses in its foundation. What happens if the people intend to take away the wealth of the prosperous? Clearly that would be unjust. For Aristotle, the answer was to reduce inequality in society and reduce the people's desire for usurping the property of the wealthy; for James Madison, the answer was to limit democracy, that is, reduce the populace's ability to exert its will on government.

From our nation's inception, powerful mechanisms were established to reduce democracy. Principal among these were the qualifications for the right to vote. In 1776, voting was controlled by state legislatures, and only White men aged twenty-one and older and who owned land were allowed to vote. In 1870, nearly one hundred years after the country's birth, the Fifteenth Amendment gave men of other races the right to vote, with the irony of excluding the nation's original inhabitants, Native Americans. Though Blacks now had the right to vote, poll taxes, literacy tests, fraud, and intimidation still prevented many from exercising their newfound right.

In 1920, women gained the right to vote when the Nineteenth Amendment was ratified, 144 years after the inception of the United States. In 1924, Native Americans were finally given the right to vote. In 1965, the federal Voting Rights Act suspended literacy tests. Registration and voting rights were then federally enforced. In 1971, the Twenty-Sixth Amendment to the US Constitution lowered the voting age to eighteen. In 1975, the Voting Rights Act was renewed, permanently banning literacy tests nationwide. Section 203 was also included in the Act, requiring translated voting materials in areas with large numbers of citizens with limited English skills.

The impetus for the 1965 Voting Rights Act was a need to protect voters, principally Black voters in the South, who'd been the victims of recurrent acts of voter intimidation and violence. While history often remembers the victors, consider for moment what can be inferred from the opposition to these egalitarian changes and the length of time it took for these amendments and laws to be passed. Clearly there were, and still are, powerful individuals and organizations that resisted the democratization of American society for their own selfish interests.

Inequality has powerful destructive consequences for society. Humans are endowed with a voracious appetite for fairness. When inequality is allowed to persist in society, an undercurrent of discontentment cannot be restrained and will manifest in a nation's populace. Drug abuse, alcoholism, hypersexual behavior, violence, protests, and crime are common ways the disenfranchised within society seek remedy for both perceived and actual wrongs.

In practical ways, the constitution was deliberately designed to reduce democracy in the wealthy's interest, further maintaining control over the populace. We see this manifested in various ways throughout American society. Initially, only White men exclusively were not only allowed to vote but also to serve as senators. They were not elected: they were chosen. What do you think were the practical consequences of having an electorate handpicked by those already in power? Certainly those chosen would likely be no threat to the established aristocratic order. The Electoral College, though described as a measure to bring balance and fairness, is simply another entrenched mechanism of subverting the will of the populace, and no other democratic society uses such a mechanism clearly designed to restrict democracy. We saw this in the election of Trump, a president who won the election, though the clear majority of Americans voted against him. He lost the popular vote by several million votes. In more recent times, gerrymandering by both the political left and right, and voting laws changes principally instigated in republican state houses, are simply new age attempts to systematically disenfranchise voters, reducing democracy in its practical effect.

Throughout its history, such has been the pattern of the United States, a nation whose stated ideals repeatedly fail to materialize in the practical application of its laws. But is this not to be expected for a country whose very foundation is rooted in the brutality of slavery? In many ways, modern American society is just the reflection of the ongoing intentional control of the masses in the interest of a controlling opulent few.

Such is not a failure of American democracy alone. It has been the pattern of every form of government for a small few to manipulate and control the government's course in their favor. Socialism is no better. In theory, it appears more egalitarian, but at its core, the practical application of socialist theory has repeatedly failed as humanity's imperfection and greed cannot be restrained.

In theory, the government is meant to protect the masses against the ever-present threat of abuse and to provide security for humanity to pursue its higher self, to engage in endeavors that satisfy the yearnings of the spirit, and fulfill the practical need to provide physically for ourselves and our families. Yet despite such universal aims and the government's awareness of these needs, we've seen throughout history, and again in more recent times, the rebuke of unjust governmental rule throughout the world. Such is the consequence of the abuse and neglect from the very entities designed and implemented for nations' protection and security.

The root of such failure lies in the lack of nobility among the rulers of humankind. When the institutions designed to provide security for humanity are inflicted with the same self-interested individuals that they're designed to provide protection from, it's no surprise the masses of humanity would feel the need to rebel against the established political class. Despite the obvious character flaws of some populist leaders today, people are so disheartened with the polished, established, polite yet utterly corrupt political establishment that they're willing to support grossly flawed individuals in the hope they will fulfill promises to give attention to the forgotten and unheard among humanity.

The Ignorance of Building Walls

Much is made of the importance of national borders. In the United States recently, a significant portion of the campaign success of Donald Trump rested on his promise of "building a wall." It was a message that resonated with many rural Whites in areas of failing economic growth throughout the United States, areas previously dependent on manufacturing jobs to raise the standard of living for less-educated, primarily working-class White men. The idea of a physical barrier to keep out individuals perceived as taking American jobs resonated deeply in these areas. Trump used that message to galvanize a wave of rural support that led him to victory through the American Electoral College, despite losing the popular vote by several million votes. The furor revealed in Trump's election shows another common fallacy present today, yet another manifestation of the internal quest for superiority that individuals routinely struggle to achieve.

Though nationalism has its roots in physical borders, those borders are also accompanied by variations in culture, language, and customs in their respective locales. Though most nations comprise individuals with a variety of ethnic and cultural backgrounds, there are national identities that geographic areas and cities within those nations adopt. We see this manifested on the microcosmic levels of towns and on the macrocosmic scale of cities and states. Even within countries, states, cities, and towns, the ever-pressing quest for superiority pits citizens against citizens. But national bonds often can unite even the most bitter divisions among citizens of the same country. During World War II and the Vietnam War, race relations were among the most divisive in US history. Yet soldiers from different backgrounds united to fight against a common enemy. A shared need for survival drove such a bond and reflected the power of government and common nationalistic bonds to overcome personal divisions.

As yet, governments have failed humanity. Society's evolution dictates humanity's survival will be achieved only through a government without a nationalistic boundary, that is, one world government. The full development and achievement of humankind's potential dictates such. Though such government is unlikely to arise through nobility of spirit, environmental catastrophe, the threat of nuclear war, or plague may force humankind's cooperation. It seems only the threat of humanity's extinction will be sufficient motivation to abandon the futility of ignorant artificial boundaries. We consider this idea of one global government further in the next chapter.

Chapter Eleven
The Inevitability of One Global Government

A revolution is coming: a revolution which will be peaceful if we are wise enough, compassionate if we care enough, successful if we are fortunate enough—but a revolution which is coming whether we wish it or not. We can affect its character, we cannot affect its inevitability.

—Robert F. Kennedy

A New World Order is in the making, and it is up to us to prepare ourselves that we may take our rightful place in it.

—Malcom X

The United Nations was born from a secret meeting between FDR and Winston Churchill in 1941. The two men met in Newfoundland and discussed the formation of an international peacekeeping organization with objectives that included postwar planning for improving world trade, disarmament, government self-determination, and social improvements for the whole of humanity. The two composed a document outlining their ideas, and that document, the Atlanta Charter, became the foundation of the Manifesto to the United Nations. Within six months, the Allies had signed the document, and by 1941 a host of other nations had joined the agreement.

The UN has succeeded in a myriad of endeavors, primarily peacekeeping efforts, protecting refugees, maintaining the integrity of elections, and a host of other meaningful and consequential initiatives. As the environment has continued to decay, the UN's mandate has broadened to include environmental protection efforts. It indeed has succeeded in many ways, and shown the flexibility to fit a wide range of international and global concerns. Still, it has not been an organization without its share of criticisms. In 2014, after Russia annexed Crimea, multiple nations expressed their displeasure over the action, but nothing impactful was done to alter the aggressor nation's course. Without the teeth to impose significant consequences on Russia, and the lack of other nations' interest and commitment to affect militarily the course of Crimea's annexation, the occupation was left unchallenged. This less-than-ideal outcome reflects the most significant deficiency of the UN. It lacks an independent and powerful military to enforce its peacekeeping efforts when a noncompliant nation is a nuclear power.

Civilization has followed a recurring pattern since its inception. Man has dominated man, to its injury. Repeatedly, the will of the populace, particularly its will for liberty and equality, has been subverted for the will of an elite few. The atrocities of war incline the masses of humanity toward instincts of peace. Yet what we see around us is not the proliferation of peace but recurring cycles of endless war in every generation. The inclination of humanity toward war is not the natural inclination of individuals as it is often espoused, as if humans are genetically inclined to large-scale warfare. The inclination of humanity is predominantly for material security and peace. The need to have a home and essential provisions for sustenance and well-being, such as medical care, are the desires that fuel humanity's quest for education and employment. War is most commonly the fruitage of ambitious and greedy men, whose insecurity drives them on a quest for superiority and fulfillment by the satisfaction derived from power and pleasure derived from what they can take. The lack of basic needs, and the lust of greed and quests for power, fuels humanity's numerous conflicts.

Conflicts, often impelled by the trivial, artificial boundaries representing no more than imaginary lines on a map, add to the tragedy of needless conflict fueled by artifice. These trivial lines, made at the auspices of men long dead, fuel ongoing disagreements between nations and groups of peoples whose ethnic, religious, and geographic identity adds fuel for division and conflict. This sad truth highlights the depth of civilization's inadequacy and the grievous state of our existence as a species. These conflicts often escalate the destruction of the natural world that surrounds us and unquestionably limit our species' effectiveness to manifest our grandest potential, where the needs of humanity and the natural world are met and harmoniously intertwined.

It is not infrequent that individual small nations make advancements toward improvements in human relations and environmental efforts. Such efforts at times can put countries separated by artificial boundaries at odds with one another. For example, one nation's water conservation efforts may negatively affect a neighboring country's agricultural interests. Were the nations not divided, accommodations could be made to achieve both ends; instead, an artificial division exists, augmenting the fuel for needless conflict.

The existence of humanity's lack of unity under one government reflects the most significant manner in which our species and civilization have failed to meet their potential. A recognition of our shared desires and needs as individuals should supersede any divisions within our species. A unified world government is the only way in which the enormous complexities and challenges to humanity's survival and harmonious relationship with the natural world can be achieved.

The Establishment of One Global Government

The establishment of one world government is inevitable. It will take only one of two possible courses: (1) willful acceptance on the part of humanity from a recognition of our species' need for unity for survival and optimization of humanity's potential, or (2) coerced establishment of one world government by the will of the masters of mankind.

Of the two, the course that allows for more conscientious planning, with attention given to universal prosperity, would be the path more suited for humanity. Still, power is never relinquished without struggle, and no degree of public sentiment in favor of such government will impel the masters of mankind to release control of humanity's resources and commerce. The only path reason dictates will manifest is that of coerced acceptance. This forced acceptance and establishment of one world government will be achieved by one of three means: (1) physical coercion from the threat of violence from a military power, (2) commercial coercion by withholding access to global trade, or (3) coercion through the need for rationing and sharing of resources as the only course for survival because of global catastrophe.

Physical coercion

In centuries past, various nations took their place as the major world powers of their time. Egypt, Mido-Persia, Greece, the Roman Empire, Barbarians, Spain, the British Empire, and America have all held the mantle of their time's dominant world powers, each fueled by a combination of military, economic, and strategic skill. These empires and others dominated large segments of the world, but none could fully conquer the world. Part of this failure was related to the global knowledge and technology of their time. For example, the Mongol Empire stretched across a vast territory, but the Mongols had no efficient transportation access to the rest of the world, so their lack of technological capability limited the breadth of their influence.

As the world advanced technologically, the global sharing of information and knowledge reduced the inequity of technology between nations. The possibility of any nation dominating the globe through sheer military might became nearly impossible. For example, multiple countries' advent of nuclear weapons has reduced the likelihood of world war on the scale previously seen since such confrontation would lead to catastrophic destruction to civilization. The chance to achieve a unified world government through one nation's physical dominance is unlikely to change as long as various nations' technological development remains roughly equivalent.

Suppose some nation were to develop some technology that vastly strengthened its position militarily, such as through a biological weapon, and weakened the threat of nuclear retaliation. In that case, that nation could theoretically accomplish the means of coercion and submission of other countries. A biological weapon of catastrophic morbidity and virulence to which an antidote was effective and access to such was controllable would provide such a tool. Given that the controlling nation's aggression in using it and controlling its application occurred surreptitiously, it would be nearly impossible to cause catastrophic loss to other countries without it becoming apparent one country was suffering considerably less morbidity from the agent, for such would immediately arouse suspicion. Though such a weapon would less likely be attainable, its dispersal and the application of an antidote to the part of the populace intended to survive would be challenging to accomplish without a significant opposing nuclear-capable nation's awareness and desire for retaliation. Difficult but not impossible though it may be, such an occurrence has a low likelihood of taking place given the nature of how logistically difficult it would be to achieve the orchestration of such a feat.

There is a segment among the masters of mankind who encourage depopulation measures as a means to confront the problems humanity faces, principally environmental decay and climate change. But depopulation measures are inherently callous and disregarding of human life, for who should have the right to decide who lives and who dies? Such measures alone would surely principally serve the interests of maintaining current power and wealth structures and would simply be another egregious example of the masters of mankind's violation of others' right to life and liberty. Such efforts, orchestrated at the hand of the elite, would certainly not occur indiscriminately, and exploitation and wanton destruction of life would simply follow the pattern of the countless egregious acts of genocide and slavery that have occurred throughout history, but in a significantly more grotesque way.

Commercial coercion

In modern times, a number of international corporate conglomerates could provide the foundation for a new global form of government. As corporations take increasing control over the distribution of goods, and food in particular, their sphere of influence becomes more entrenched in society. The management of goods and food provides both the means and opportunity for various ways of affecting the world's population. As companies merge and larger segments of commerce, food distribution, and collected private digital information as a commodity are further concentrated, these companies' power and spheres of influence will supersede even powerful national governments.

However, the only thing keeping national governments' power in place is their military strength. In modern times, military technology has increasingly become privatized, and the power of military contractors has substantially increased. Dependence on these suppliers will make governments increasingly susceptible to exploitation, manipulation, and control. As data is increasingly collected through monitoring of phone and computer devices, the ability to manipulate and control the population is enhanced, as is the ability to identify dissidents and others within the population likely to resist such changes.

Such control by global corporations in time could become so immense that the near-complete control of the worldwide economy will rest in the hands of a very select few. The privatization of military technology also increasingly rests in the hands of a few companies, which are not unlike other industries and corporations and are subject to be bought and sold to the highest bidder. A superwealthy corporation, with control over not just trade and food but also military technology, could easily exert its will over even immensely powerful governments. Corporate power could additionally be easily reinforced by an arsenal of trained private human soldiers. The development of AI and robotics will make even human soldiers inconsequential in time.

Coercion as means of survival

As displayed throughout humanity's existence, there is no limit to the brutality, manipulation, selfishness, and ignorance society as a whole is capable of displaying. While the ambition of corporations and governments to achieve domination of the globe knows no bounds, nature has a way of having its say in the course of Earth's destiny. The coronavirus has shown the power of the natural world to alter the course of humanity. An increasingly unstable climate and the ever-present threat of emerging infectious diseases reveal that humanity's course can be altered in an instant, and the possibility of the natural world threatening the continued existence of our species is an ever-present danger.

The natural world is fragile, and humanity itself has become its most significant parasite. One of my favorite Bob Marley songs, "I Shot the Sheriff" has some lyrics whose profundity extends far beyond human relations.

> Every day the bucket a-go a well,
> One day the bottom a-go drop out.

Those lyrics were not of Marley's originality, for they were born from a popular Jamaican idiom. I'm reasonably sure they were not meant to apply to our current global climate condition, but they indeed depict well the global crisis humanity now faces. For centuries, humankind has taken the bucket to the well, developing technology without considering the environmental ramifications and abusing the natural world by polluting it and exploiting its organic and inorganic resources. And we are watching "the bottom a-go drop out" as our climate grows more and more unstable by the day.

The world has been watching, mostly indolently, as our planet reaches an apex of catastrophe, and humanity will not indefinitely be able to ignore what has been evident for decades. Rising sea levels, the product of glaciers melting gradually, will no longer be able to be denied as coastal cities experience irrevocable flooding. But by then, the cure for our environment's troubling decline will not reverse centuries of injury instantly.

The subsequent displacement of the populace of these cities and the food shortages, disease, and violent conflicts that will spread as a result of our steadily worsening environment and climate will produce a degree of suffering and despair unique in humanity's history. Humankind is beyond the point of rescuing our world from what will be, not because we like the capability to do so but because we lack the structure within our governments and the will within the populace to significantly amend our addiction to consumerism. We lack the ability to resist the powerful influence of the masters of mankind, whose insatiable thirst for wealth at the expense of humanity's welfare will not be quelled.

It is estimated that by 2100, at least eleven major cities worldwide will be so flooded they will essentially be underwater and uninhabitable in the way we now know them. These include Miami, Florida (470,000 people); Jakarta, Indonesia (9.6 million people); Alexandria, Egypt (5.3 million people); Virginia Beach, Virginia (1.5 million people); Bangkok, Thailand (10.5 million people); New Orleans, Louisiana (400,000 people); Rotterdam, the Netherlands (650,000 people); Venice, Italy (640,000 people); Dhaka, Bangladesh (21 million people); Houston, Texas (7 million people); and Lagos, Nigeria (14.4 million people). These are a relatively small number of cities, but their combined population numbers approach nearly 75 million today and could easily approach more than 100 million by 2100.

And they are not the only cities that will suffer the effects of our increasingly unstable climate. Rising global temperatures will have significant deleterious effects on both agricultural production and biodiversity. I don't pretend to be an expert on climate change, nor do I fully understand all the ways an increasingly unstable climate will affect humanity. But I don't think it takes an expert to comprehend that humanity's complex climate challenges are daunting and significant, and I do know that confronting these challenges will take a concerted global effort, an effort that our civilization, as of yet, has failed to rise to the level of determination and sacrifice needed to redress.

But climate instability is not the only means by which nature effects enormous change upon humanity. In late 2019, and extending throughout 2020, the world saw the ravages of what a microscopic infectious agent can do on a relatively small scale. COVID-19's mortality rate was mostly dependent on age, with seniors above age eighty-five manifesting the highest death rate at nearly 15 percent. Infants and young adults were extremely unlikely to die from the infection, yet worldwide, COVID-19 has led to almost 5 million deaths as I write these words. While 5 million deaths are a small portion of the world's population, they represent an unquantifiable value in the loss of human life. The economic loss to working-class people has been catastrophic, while large corporations like Amazon and Walmart, and financial institutions prospered considerably. To date, Amazon has hired more than 175,000 employees since COVID-19 began, while Walmart has hired 400,000. The period during COVID-19 has comprised a significant consolidation of wealth and land for financial institutions, as struggling small businesses and homeowners suffered losses in the face of the pandemic's economic ramifications.

COVID-19's overall mortality has been low, near 1 percent, but it gives us a glimpse of how quickly the world can change in the face of a global pandemic. Had COVID-19's mortality been 10 to 15 percent or worse across all ages, imagine the carnage and economic impact. The virus is still undergoing significant mutation. To date, the mutations have predominantly increased the ability of the virus to spread. Should a mutation in COVID-19 result in a significant increase in its mortality, 25 to 50 percent perhaps, the world would be crippled in a way no plague has ever affected human civilization. There is no reason to believe such a mutation is inevitable, as far as I know. Still, it's a possibility, and the possibility of another pandemic with the potential to mimic COVID-19's highly contagious nature, combined with a significant mortality rate, would force humanity to deal with such a plague in a concerted, unified way.

I'm not of the mind-set that humanity will willfully embrace one world government. I also have little confidence our species will avoid the inevitable consequences of global climate change. The pattern of global consumerism and the pressure to maintain such lifestyles have become ingrained in society under the influence of the masters of mankind, and such patterns will be impossible to break without the coercion of circumstance. Though I sincerely hope I'm wrong in such an assessment, I feel far more confident the growing uncertainty of our species' survival, under threat from the global climate crisis; concerted economic coercion; or unprecedented pandemic threat will eventually lead to the formation of one world government. In as much as a blessing in disguise can blossom from the seeds of an unwanted and difficult circumstance, perhaps humanity's survival will blossom from catastrophe, and it appears, at least at present, our species is inevitably destined for such.

The Bible and One World Government

Over the years I've developed a great fondness for world religion. Perhaps born from a need to find a deeper meaning in life and suffering, I've considered nearly every organized religion worldwide and read the holy books associated with them. For me, I think partly from my cultural heritage and familial ties, the Bible and Buddhist teachings have always ended up being the philosophies I've found most fascinating and thought provoking, irrespective of their religious significance. I have read the Bible, specifically, no fewer than at least a dozen times.

I always like to get the sense of things on my own, and particularly so in reading the Bible. Devoid of any religious indoctrination or established religious customs, the Bible offers an interesting take on one world government that is often overlooked.

The Bible isn't a book we usually think of as having a theme, but it does indeed have a theme that permeates its pages, from the so-called Old Testament to Revelation, and I believe that theme is the coming of one world government. When I read the Bible, I've always felt a fondness for the books of Job, Proverbs, Ecclesiastes, and Psalms and for the Sermon on the Mount. These texts pose beautiful and complex philosophical questions and stylistically feature everything from beautiful poetry to the masterpiece of a monologue attributed to Jesus in the Sermon on the Mount. I appreciate the beautiful aspects of all religions insofar as they encourage moral and benevolent conduct. But a closer look at the Bible reveals a subtle, well-hidden theme that is often left unappreciated even by many Christians themselves.

In this account in Daniel 2, Daniel writes about a period in his life when King Nebuchadnezzar had a dream that troubled him so much he summoned all the astrologers, magicians, enchanters, and sorcerers in his kingdom. He tasked them not merely with interpreting the dream, but to test them, they had to tell him the details of the dream without him revealing it. It was a task that would unquestionably establish who among them truly had clairvoyant ability. Consider the words of the account below.

Nebuchadnezzar's Dream

[1] In the second year of his reign, Nebuchadnezzar had dreams; his mind was troubled and he could not sleep. [2] So the king summoned the magicians, enchanters, sorcerers and astrologers to tell him what he had dreamed. When they came in and stood before the king, [3] he said to them, "I have had a dream that troubles me and I want to know what it means."

[4] Then the astrologers answered the king, "May the king live forever! Tell your servants the dream, and we will interpret it."

⁵ The king replied to the astrologers, "This is what I have firmly decided: If you do not tell me what my dream was and interpret it, I will have you cut into pieces and your houses turned into piles of rubble. ⁶ But if you tell me the dream and explain it, you will receive from me gifts and rewards and great honor. So tell me the dream and interpret it for me."

⁷ Once more they replied, "Let the king tell his servants the dream, and we will interpret it."

⁸ Then the king answered, "I am certain that you are trying to gain time, because you realize that this is what I have firmly decided: ⁹ If you do not tell me the dream, there is only one penalty for you. You have conspired to tell me misleading and wicked things, hoping the situation will change. So then, tell me the dream, and I will know that you can interpret it for me."

¹⁰ The astrologers answered the king, "There is no one on earth who can do what the king asks! No king, however great and mighty, has ever asked such a thing of any magician or enchanter or astrologer. ¹¹ What the king asks is too difficult. No one can reveal it to the king except the gods, and they do not live among humans."

¹² This made the king so angry and furious that he ordered the execution of all the wise men of Babylon. ¹³ So the decree was issued to put the wise men to death, and men were sent to look for Daniel and his friends to put them to death.

¹⁴ When Arioch, the commander of the king's guard, had gone out to put to death the wise men of Babylon, Daniel spoke to him with wisdom and tact. ¹⁵ He asked the king's officer, "Why did the king issue such a harsh decree?" Arioch then explained the matter to Daniel.

[16] At this, Daniel went in to the king and asked for time, so that he might interpret the dream for him.

[17] Then Daniel returned to his house and explained the matter to his friends Hananiah, Mishael and Azariah. [18] He urged them to plead for mercy from the God of heaven concerning this mystery, so that he and his friends might not be executed with the rest of the wise men of Babylon. [19] During the night the mystery was revealed to Daniel in a vision. Then Daniel praised the God of heaven [20] and said:

"Praise be to the name of God for ever and ever;
    wisdom and power are his.
[21] He changes times and seasons;
    he deposes kings and raises up others.
He gives wisdom to the wise
    and knowledge to the discerning.
[22] He reveals deep and hidden things;
    he knows what lies in darkness,
    and light dwells with him.
[23] I thank and praise you, God of my ancestors:
    You have given me wisdom and power,
you have made known to me what we asked of you,
    you have made known to us the dream of the king."

Daniel Interprets the Dream

[24] Then Daniel went to Arioch, whom the king had appointed to execute the wise men of Babylon, and said to him, "Do not execute the wise men of Babylon. Take me to the king, and I will interpret his dream for him."

[25] Arioch took Daniel to the king at once and said, "I have found a man among the exiles from Judah who can tell the king what his dream means."

26 The king asked Daniel (also called Belteshazzar), "Are you able to tell me what I saw in my dream and interpret it?"

27 Daniel replied, "No wise man, enchanter, magician or diviner can explain to the king the mystery he has asked about, 28 but there is a God in heaven who reveals mysteries. He has shown King Nebuchadnezzar what will happen in days to come. Your dream and the visions that passed through your mind as you were lying in bed are these:

29 "As Your Majesty was lying there, your mind turned to things to come, and the revealer of mysteries showed you what is going to happen. 30 As for me, this mystery has been revealed to me, not because I have greater wisdom than anyone else alive, but so that Your Majesty may know the interpretation and that you may understand what went through your mind.

31 "Your Majesty looked, and there before you stood a large statue—an enormous, dazzling statue, awesome in appearance. 32 The head of the statue was made of pure gold, its chest and arms of silver, its belly and thighs of bronze, 33 its legs of iron, its feet partly of iron and partly of baked clay. 34 While you were watching, a rock was cut out, but not by human hands. It struck the statue on its feet of iron and clay and smashed them. 35 Then the iron, the clay, the bronze, the silver and the gold were all broken to pieces and became like chaff on a threshing floor in the summer. The wind swept them away without leaving a trace. But the rock that struck the statue became a huge mountain and filled the whole earth.

36 "This was the dream, and now we will interpret it to the king. 37 Your Majesty, you are the king of kings. The God of heaven has given you dominion and power and

might and glory; 38 in your hands he has placed all mankind and the beasts of the field and the birds in the sky. Wherever they live, he has made you ruler over them all. You are that head of gold.

39 "After you, another kingdom will arise, inferior to yours. Next, a third kingdom, one of bronze, will rule over the whole earth. 40 Finally, there will be a fourth kingdom, strong as iron—for iron breaks and smashes everything—and as iron breaks things to pieces, so it will crush and break all the others. 41 Just as you saw that the feet and toes were partly of baked clay and partly of iron, so this will be a divided kingdom; yet it will have some of the strength of iron in it, even as you saw iron mixed with clay. 42 As the toes were partly iron and partly clay, so this kingdom will be partly strong and partly brittle. 43 And just as you saw the iron mixed with baked clay, so the people will be a mixture and will not remain united, any more than iron mixes with clay.

44 "In the time of those kings, the God of heaven will set up a kingdom that will never be destroyed, nor will it be left to another people. It will crush all those kingdoms and bring them to an end, but it will itself endure forever. 45 This is the meaning of the vision of the rock cut out of a mountain, but not by human hands—a rock that broke the iron, the bronze, the clay, the silver and the gold to pieces.

"The great God has shown the king what will take place in the future. The dream is true and its interpretation is trustworthy."

46 Then King Nebuchadnezzar fell prostrate before Daniel and paid him honor and ordered that an offering and incense be presented to him. 47 The king said to Daniel, "Surely your God is the God of gods and

the Lord of kings and a revealer of mysteries, for you were able to reveal this mystery."

[48] Then the king placed Daniel in a high position and lavished many gifts on him. He made him ruler over the entire province of Babylon and placed him in charge of all its wise men. [49] Moreover, at Daniel's request the king appointed Shadrach, Meshach and Abednego administrators over the province of Babylon, while Daniel himself remained at the royal court.

—Daniel 2

Daniel describes the coming of global government

The Bible states Daniel not only interprets but also describes the king's dream in detail, without having any prior interaction with Nebuchadnezzar. He describes a history of kingdoms beginning with the kingdom of Nebuchadnezzar himself. The dream culminates in the description of a government in the future that will be the last and that will rule over Earth forever. He describes it in Daniel 2:44:

> In the time of those kings, the God of heaven will set up a kingdom that will never be destroyed, nor will it be left to another people. It will crush all those kingdoms and bring them to an end, but it will itself endure forever.

This same theme is repeated throughout the Bible and featured in the model prayer that Jesus uttered during his most famous sermon, the Sermon on the Mount:

> This, then, is how you should pray: "Our Father in heaven, hallowed be your name, your kingdom come, your will be done, on earth as it is in heaven."
> —Matthew 6:9

> But he said, "I must proclaim the good news of the kingdom of God to the other towns also, because that is why I was sent."
> —Luke 4:43

Many Christians often repeat the model prayer, not fully appreciating its connection with the kingdom described in Daniel. It is significant to consider that the king of this kingdom, whose lineage is described in a number of promises made to well-known biblical figures, including Abraham, Isaac, Jacob, and Judah, makes up a significant portion of the Old Testament.

Consider what Bible writers say about the lineage of the coming king, a king who Jews believe is yet to be identified but who Christians believe is no other than Jesus himself. The promise was made to Abraham that God would turn his offspring into a nation that included kings. That same promise was made to Abraham's son Issacs and his son Jacob. The promise of the king was then said to come through the line of Judah, and next narrowed to come through the line of King David.

While most Christians focus on a reward of heavenly life, some scriptures highlight that the fruitage of that kingdom will affect life on Earth to a significant degree, as highlighted in part by Jesus's words in the model prayer. "Let your kingdom come, let your will be done on Earth." Would not the effect of God's will being done on Earth result in some change on Earth from a Christian's point of view?

The book of Isaiah is a book full of prophecy, according to both Jews and Christians, and its words highlight what some believe are the conditions that will exist on Earth as a result of the one world government to be established by the God of both Jews and Christians. Consider some of the conditions described in the book of Isaiah:

> No one living in Zion will say, "I am ill"; and the sins of those who dwell there will be forgiven.
> —Isaiah 33:24

> "They will build houses and dwell in them; they will plant vineyards and eat their fruit."
> —Isaiah 65:21

> 6 The wolf will live with the lamb,
> the leopard will lie down with the goat,
> the calf and the lion and the yearling together;
> and a little child will lead them.
> 7 The cow will feed with the bear,
> their young will lie down together,
> and the lion will eat straw like the ox.
> 8 The infant will play near the cobra's den,
> and the young child will put its hand into the viper's nest.
> 9 They will neither harm nor destroy
> on all my holy mountain, for the earth will be filled with the knowledge of the Lord as the waters cover the sea.
> Isaiah 11:6–9

> He will judge between the nations
> and will settle disputes for many peoples.
> They will beat their swords into plowshares

and their spears into pruning hooks.
Nation will not take up sword against nation,
nor will they train for war anymore.

—Isaiah 2:4

For Christians, the Bible describes conditions that will exist on Earth among humanity before the one world government's coming in the New Testament, as 2 Timothy 3:1–5 illustrates:

> [1] But mark this: There will be terrible times in the last days. [2] People will be lovers of themselves, lovers of money, boastful, proud, abusive, disobedient to their parents, ungrateful, unholy, [3] without love, unforgiving, slanderous, without self-control, brutal, not lovers of the good, [4] treacherous, rash, conceited, lovers of pleasure rather than lovers of God— [5] having a form of godliness but denying its power. Have nothing to do with such people.
>
> —2 Timothy 3:1–5

Revelation 11:18 described the ruining of the Earth when pollution was not even a concept of the writer, the apostle John:

> The nations were angry, and your wrath has come. The time has come for judging the dead, and for rewarding your servants the prophets and your people who revere your name, both great and small—and for destroying those who destroy the earth.
> —Revelation 11:18

Jesus himself utters the last sign that precedes the coming of the kingdom described throughout the Bible—and the very last sign:

> "And this gospel of the kingdom will be preached in the whole world as a testimony to all nations, and then the end will come."
>
> —Matthew 24:14

Of significance, in relation to those words, Jesus uttered another prophesy:

> [21] "For then there will be great distress, unequaled from the beginning of the world until now—and never to be equaled again.
>
> [22] "If those days had not been cut short, no one would survive, but for the sake of the elect those days will be shortened."
>
> —Matthew 24:21-22

Whether these statements portend to a pandemic or global physical catastrophe, it is not specified. But the words are ominous nonetheless. I find it intriguing the Bible portends environmental decay and environmental catastrophe, despite being written thousands of years ago. Many will argue that such phenomena have always occurred throughout the history of humanity's presence; however, there is a prophecy uttered by Jesus that I find difficult to explain away: "And this good news of the kingdom will be preached in all the inhabited Earth, and then the end will come."

When Jesus uttered those words, he was just one man preaching. There were no newspapers, televisions, radios, or social media. And yet he predicted, accurately, that his message would spread throughout Earth. It is a realization worth considering, and it lends credence to considering Jesus was possibly no ordinary person. When

Jesus uttered the words regarding the message of the kingdom, one world government, the only effective form of communication was through letters and word of mouth. It is a significant development for which I cannot find a logical explanation.

The Ruler of the Global World Government

Both democracy and socialism are beautiful forms of government in theory. Still, the frailty of inherent human weakness and humanity's propensity for greed and selfishness turn both of these forms of government into failures for humanity in practice. For democracy, the illusion of equality masks the actual function of democracy in practice to preserve the rich's wealth and land through the intentional limiting of equality and restricting of democracy and power. For socialism, indoctrination and restrictions of freedoms are used to compel the populace in socialist nations to accept a manner of life with considerably fewer luxuries than are present in countries of more wealth, often the results of consumerism and capitalism, which flourish in democratic nations. It has also been the pattern of socialist rulers to profit personally from the facade of equality, while the populace's labor and resources are used to increase the personal wealth of the rulers themselves. Fidel Castro is the quintessential example of this pattern.

Even when democratic and socialist nations know the rulership of benevolent and competent rulers, rulers dedicated to egalitarian principles and genuine concern for the entirety of the people they govern, their rulership is limited. Restricted either by mortality, constitutional limits on their terms, or by opposing segments within the political arena who lack such concern for the welfare of the masses, a single ruler and the benefits of their benevolent rulership rarely extended beyond the lifespan of that ruler. In democracy's current state in America, this is ever more pertinent as corporations exert increasingly broad political influence as the cost of political elections increases. To remain competitive in political races, politicians are repeatedly pushed into the hands of wealthy corporations—corporations whose interests inherently place politicians against the interests of the working class. Corporations vehemently fight to keep worker wages low and unions' power and liberties weakened to maximize corporate profits for a select few elites.

The UN, with all its noble accomplishments, and there have unquestionably been many, has failed dramatically in one very significant arena. It has failed to take a stand on perhaps the most significant issue affecting the very *maxime amet* it claims to pursue. The improvement of society is inherently interconnected with the reduction of inequality of the global populace, and every significant measure of well-being within humanity is inextricably tied to wealth and social class. Health, both mental and physical; generational cycles of poverty and wealth; land ownership; educational opportunities; gender equality; and many other societal problems are all intricately and fundamentally tied to poverty and, particularly, cycles of poverty and generational wealth.

This is not some unique revelation. Social scientists deeply understand the connection between poverty, economic inequality, and social justice, which undoubtedly includes equality in all avenues of society. And so it begs the question, why has the UN never taken a public stance against poverty? What would such a posture even look like? Noam Chomsky, an intellectual who understands the dynamics of politics and wealth in an elegant and nuanced way, describes in great detail the period of financialization of the American economy in the 1970s through today. This period marked the simultaneous decline in American manufacturing and the unabated growth of American financial institutions.

As a rule, capital is much freer to move than labor. For example, workers, especially working-class laborers, are restricted in their ability to move geographically for a variety of reasons, while companies can freely move money anywhere in the world they want to, for the most part. As a result, the outsourcing of American manufacturing to countries that grossly exploit workers leads to increased corporate profits and growing inequality around the world. This growing inequality has marked consequences for society, and without the political power to stem policies in opposition to the will of the masters of mankind, this pattern continues today. It has expanded to include not just manufacturing but also the sale of goods by major internet retailers like Amazon and Walmart, who represent the latest evolution and iteration of this phenomenon. They can acquire goods at the lowest possible price. The convenience of mail delivery and the comparable cost savings of purchasing their products over brick-and-mortar stores squeezes main street businesses that lack the demand, capital, and influence to acquire the same goods at comparable prices.

We're headed to a world, and some would argue we're already there, where an assortment of a select few corporations control virtually every sector of trade, superseding even the power of governmental superpowers. The product of decades of corporate mergers, these corporations hold mammoth international influence, and driven solely by the bottom line, the cost to the working class and the subsequent growing inequality that results is sure to grow without significant intervention. The time for the UN and governments to address poverty is riper than ever. But will they? And even if they do, would the UN even have the influence and power to make a significant difference? During the latest G7 summit, a decision was reached to establish a minimum tax rate for corporations. While such an effort is directed to stem the increasing power of corporations to move capital, such effort speaks nothing of these governments' efforts to protect workers and reduce inequality. It is governmental effort purely to strengthen their own economic power and reduce the power of corporations to freely move operations to nations with more favorable tax incentives.

The obvious answer, and the one that will undoubtedly be most controversial, is that an international labor union should be established, as well as an international minimum wage standard. There is no doubt that such a union and wage standard will raise the ire of the masters of mankind, as both wealthy and poorer nations where corruption abounds unabatedly would object to such an organization. It would be considerably challenging for the UN to maintain its neutrality in international affairs while advocating policies that could have significant economic effects, more on some nations than others. There is every reason to believe that such a union, with the directive aimed at increasing the standard of living for the working class with a wage commensurate with more developed industrialized nations, would not stagger the economies of smaller, less affluent nations but would instead stimulate them, as the poor and working class must spend their earnings to sustain their lives.

The Outcome of Global Government

How would a global government even function in practice? How should a leader be chosen? I find it fascinating that the Bible outlines the establishment of a global government taking root after global catastrophe threatens the survival of humanity as a species, exactly what we see playing out on the world stage at present. For Christians, Jesus represents the ideal ruler, and the accounts of Jesus in the Bible, in many ways, play as far more than myth. There is the feeling of compelling veracity in some accounts, and complete nonsense in others. Such elevates Jesus, not merely as some hero in a myth but to the stature of savior for Christians, or fictionalized baloney for others. His sacrificial death and the elements of his character, including his compassion for the poor and indigent, represent benevolent qualities in an ideal ruler, though particularly uncommon. His bravery, as displayed in the confrontation of merchants in the temple, combined with his compassion prophesied in Isaiah, where it parallels the conditions of the one world government, provide a model of leadership for humanity to aspire to.

In the context of these prophecies of one world government, the Bible paints an image of Earth where every member of society enjoys peace and prosperity. In this utopia, animals are tame, war is a thing of the past, and humanity is united with the liberty to pursue their noblest aspirations. It is a beautiful aspect of the Bible that is often left unappreciated. Christians nearly uniformly hold to an image of aspiring to heavenly life. Yet there are some salient philosophical questions that both Jews and Christians should consider when analyzing the Hebrew Scriptures. Why did their God make Earth in the first place? Did it have a purpose? What was the original mandate given to Adam?

> For this is what the Lord says—
> he who created the heavens,
> he is God;
> he who fashioned and made the earth,
>   he founded it;
> he did not create it to be empty,
>   but formed it to be inhabited—
> he says:
> "I am the Lord,
>   and there is no other."
>
> —Isaiah 45:18

Earth was made to be inhabited, and it will stand forever. Does it not stand to reason that the intention indicated in these scriptures was for humankind to live on Earth indefinitely.

> Generations come and generations go,
> but the earth remains forever.
> —Ecclesiastes 1:14

> [8] "For my thoughts are not your thoughts,
>   neither are your ways my ways,"
>   declares the Lord.
> [9] "As the heavens are higher than the earth,
>   so are my ways higher than your ways
>   and my thoughts than your thoughts.
> [10] As the rain and the snow
>   come down from heaven,
> and do not return to it
>   without watering the earth
> and making it bud and flourish,
>   so that it yields seed for the sower and bread for the eater,
> [11] so is my word that goes out from my mouth:
>   It will not return to me empty,
> but will accomplish what I desire
>   and achieve the purpose for which I sent it.
> —Isaiah 55:8–11

> Now to him who is able to do immeasurably more than all we ask or imagine, according to his power that is at work within us.
> —Ephesians 3:20

When I read these words, thinking of them void of any religious significance, they paint an extraordinary picture of a world in which humanity can reach the apex of our potential. United in peace, with prosperity enjoyed by all, such a world is the noblest aspiration humanity can hope to achieve. These words ring rich with symbolism. For the Christian and Jew, they offer a message filled with the hope and possibility only a God of immense power could bring to fruition. For those without such faith in God, the words paint in symbolic terms a world that humanity should strive for, which, in some sense, I think just as significant. One of the practical aspects of religion I find most appealing is the drive for improving interpersonal relationships. In this regard, the impetus is not necessarily about the religious literal implications of these words, but rather the drive for a fairer society.

When I read various religious texts, I'm inspired by the moral and ethical jewels they all espouse, often in very similar terms. Whether you interpret these books as inspired by God or as humanity's imagination and thirst for egalitarian social conditions, there are lovely thoughts whose expression inspire nobility. They paint for humanity the possibility of a genuinely elevated civilization. So far as I know, of all the predominant holy books, the Bible alone speaks of one world government as a hope for humanity. My work in understanding unseen entities has led me to reconsider my perceptions of the Bible and its messages in ways I had not done previously. Whether its prophecies have merits or not, the idea of one world government that allows for a measure of material prosperity for all its residents represents the noblest of all political theory. Ruled by a noble ruler, or rulers, dedicated to benevolent rulership, humanity will undoubtedly prosper. It is the only path by which our species globally will indeed live in harmony with the natural world, and the only path for achieving our highest and grandest aspirations as a civilization.

Author's note: *This chapter was written approximately six months prior to the Cuba freedom movement that fomented on July 11, 2021. My stance on the US embargo on Cuba has changed significantly after watching the millions who've taken to the streets in Cuba protesting for their freedom. I no longer wish to see the embargo ended, but I have chosen to leave the chapter in its original written form.*

Chapter Twelve
The Cuban Dilemma: Is Democracy the Answer?

To all Socialists, I wish for you: the abundance of Venezuela, the salary of Cuba, the justice of China, and the Freedom of North Korea.

—Javier Milei

Like stones rolling down hills, fair ideas reach their objectives despite all obstacles and barriers. It may be possible to speed or hinder them, but impossible to stop them.

—José Martí

The seeds of liberty and hopes of prosperity, planted in no small measure through the power of the internet, cannot be wiped from the minds and hearts of the Cuban people, no matter how much they are repressed. The only question remaining is what the manner and degree of change will be. Change is inevitable for Cuba. For decades, freedom of speech and expression have been greatly restricted on the island, as Cuban citizens lived under the threat of imprisonment just for saying something against Fidel Castro or his government. But Cuba's future now rests entirely in the hands of its people, and whatever destiny is born in Cuba, it will be driven by their wishes. The Cuban government would be wise to accept this, for change by choice is far better than change by revolution. And make no mistake, revolution will come, and come quickly, if the Cuban government does not swiftly alter is policies and ways of governance. But is the obliteration of socialism the only path forward for greater liberty and freedoms to be won by the Cuban people? Perhaps not.

The hunger for freedom is inherent in humankind, and no manner of indoctrination or repression will rip from the hearts of humanity what nature has born into our spirit. The aura of change in the air in Cuba springs from a generation without fear of or admiration for Castro. The internet has shown the Cuban people the comforts and luxuries of consumerism and capitalism, and the appetite for what others in more materially prosperous nations know as their reality cannot be quelled. This appetite has led to the first significant demonstrations of opposition on the island in decades, and though small, there is no reason to believe that such opposition will not continue to blossom and grow as the people remain committed to enjoying the freedoms and standard of living much of the rest of the world knows.

Among my many friends on the island are a collection of intellectuals and professionals, including a large number of physicians and artists. There is great discontent among them, and for good reason, but it is often left unspoken from fear. Physicians in Cuba make less money a month than an American fast-food worker makes in a day—a meager $60 a month—while other workers in the Cuban tourism industry, like taxi drivers, make considerably more. The average unskilled worker on the island makes $20 a month. Beyond that, Cuba exports physicians to other parts of world for wages similar to what US physicians earn. Those same physicians, earning Cuba thousands of dollars a month in service to other nations, receive a meager $100 to $120 a month while the Cuban government takes more than 99 percent of their wages. It is a reality that is akin to modern-day slavery, and it is contributing enormously to an air of discontent on the island. This discontent is not a feeling shared by physicians alone, but rather one proliferating throughout the entirety of the Cuban population, particularly so as a generation with no fealty to Fidel Castro comes of age and grows more familiar with social media.

Whatever your personal feelings about Castro, he was a brilliant, charismatic, and ruthlessly calculating leader. He led a revolution that overthrew the US-backed Batista—no small accomplishment, especially considering that much of Castro's force initially was comprised of untrained and poorly armed soldiers. He maintained his fledgling new government through the sheer determination of his will and subterfuge of all manner, including murder. It was a task that, irrespective of my personal feelings about the manner in which it was accomplished, illustrates the strength of his skill as a military commander and politician. But Fidel is dead, and his children did not follow his political aspirations. His brother Raul, who led the nation briefly following his death, has handed the reigns of the nation off to a new leader, and Cuba is positioned now to chart a new course, should the people and the government have the will do so. It is a reality that despite Cuba's many failings and challenges offers tremendous hope for the Cuban people and the world.

The voice of dissidents, long repressed by the Castro regime, have reinvigorated calls for democracy and freedom in Cuba. And their voices are being heard. Throughout the island, in a way that never happened during the days of Fidel, people are voicing and demonstrating their discontent with the economic, social, and political conditions that exist on the island; and they are doing it despite enormous personal consequence. Dissidents on the island face not only imprisonment, but their children and family members also face reprisals. Expulsion from school, being deemed social outcasts, ridicule, and unemployment are frequent and common forms of reprisal and repression in Cuba and have historically been enormously successful at controlling the Cuban populace, despite widespread and nearly universal poverty. But a young generation of Cubans, fearing nothing to lose in a system they perceive cannot get any worse, just don't give a fuck. When a nation's populace reaches this degree of perceived injustice and corruption in its leaders, combined with extreme poverty, it's just a matter of time before revolution is inevitable given such conditions persist.

On my first trip to Cuba, my wife's grandfather and I spent hours talking and getting to know each other. While my nights were filled with dancing, games of dominoes and chess, and emptying bottles of rum with my father-in-law and his friends around a roasting pig or goat, my days were spent talking politics and history with my wife's grandfather.

Though an elderly man of nearly eighty, he looks to be in his sixties. Slim and fit, the fruitage of walking eight flights of steps multiple times a day to and from his small three-bedroom apartment, he dresses with a neatness and dignity that befits a man of his intellectual stature. Though you wouldn't know it from his kind and approachable demeanor, he is a significant intellectual figure in Cuba. He was a young lawyer at the time of the revolution, and he grew to become one of the most influential lawyers in Cuba in the days of Fidel. He was known affectionately as the people's lawyer because of his benevolent disposition and his tireless efforts to serve the people of his community. He is also a law professor and has handled several prominent legal cases in Cuba, and he is an erudite student of history.

In the hours we spent talking, we spoke of many things. Fidel, the Cuban War of Independence with Spain, and the new Cuban constitution he recently assisted in completing were our most frequent topics of discussion. His perspective on the world and Cuban history were invaluable as I grew to understand more about a country I'd only learned about from books and conversations with my wife. My wife's grandfather is a man fond of telling stories from history, particularly the history of Cuba. When my wife was young, instead of spanking her he would tell her stories for hours when she needed disciplining. She hated it, but the stories stuck and she, too, has become an expert storyteller and writer in her own right.

One of the stories he shared that always resonated with me was that of an American soldier by the name Henry Reeve. Henry Reeve started as a drummer boy for the Union Army during the American Civil War, and at the end of the war he traveled to Cuba. In the 1860s, America supported Cuba in a war against the Spanish, who'd controlled Cuba for centuries. When Reeve exited the ship that took him and other soldiers to Cuba, several Spanish soldiers ambushed and captured them. They were lined up and shot by a Spanish firing squad. Reeve was wounded and left for dead, but he survived his wounds and crawled away until he was eventually found by a Cuban soldier.

In time, Reeve's wounds healed, and he dedicated himself to the Cuban effort to gain independence from Spain. He fought in more than four hundred battles in seven years with the Cuban Independence Army and learned Spanish by reading a copy of *Don Quixote* he found in a field. A valiant fighter, Reeve eventually became a brigadier general in the Cuban Independence Army. After suffering a severe leg injury when leaping over an artillery canon, he was never able to walk again without metal braces, but Reeve was determined to fight on. In subsequent battles, he was harnessed to his horse, and he continued to fight. He eventually lost his life in the fight for Cuba's independence from Spain but became a renowned figure in Cuba's history.

Henry Reeve was a an American. He was just one of many American soldiers who gave their lives for Cuba's independence from Spain in the late 1800s. His story illustrates one of the valuable lessons that history often teaches us: Political alliances are often flimsy, especially when solely rooted in economics. Less than a hundred years after his sacrifice on Cuba's behalf, Cuba went through another revolution. This time Fidel Castro led a fight to overthrow the regime of Fulgencio Batista, and America has opposed the communist Cuban regime Castro established ever since.

What changed? How did America go from being Cuba's ally to its enemy? Principally, the change was rooted in economics. Fidel confiscated land, including American business assets and property, and as Cuba ceased being a source of revenue for American corporations, America withdrew. With Spain now absent from Cuba, America had little military strategic interest in the island.

The political motivation of America has changed little in the time that has passed since Castro's revolution, and the Cuban government remains committed to maintaining the socialist roots established by Fidel. Part of Cuba's persistence in maintaining its independence is driven by maintaining control over its economic and political system, and necessity dictates resisting the capitalist interests of the US to solidify that aim. That is an entirely reasonable position to maintain from the Cuban government's perspective, given what history teaches us about the power of economics and land ownership to exert political influence, and how money influences politicians and governments to frequently make decisions in direct contrast to the best interests of the people they govern. This is a pattern that has repeatedly occurred in the United States' own history, with glaring examples like the Citizens United supreme court decision and the history of voting rights opposition in the US favoring affluent land and business owners.

For many, the perception of socialism and communism is skewed by the history of repression and abuses communist governments have placed on their citizenry. The actions of a ruling elite within the framework of communism are not an indictment of socialism and communism as political theories, but more so of the greed and cruelty of human nature and the corruption that power engenders. For in as much as communism's history around globe has been one of repression and exploitation of the populace for the benefit of a ruling class, democracy has been the tool for democratic nations to protect the wealth of a ruling class of elites.

In practice, communism survives on the repression and restriction of freedoms, coupled with indoctrination, while democracy thrives upon the illusion of equality and freedom. The atrocities of communism appear more egregious, as communist nations are predominantly less affluent than democratic nations and the means of repression are more overt. Democracy survives on the illusion of equal power, the fallacy of the government being an expression of the populace's will, coupled with the engrossing distraction of consumerism. In reality, that illusion keeps the populace restrained from revolution, with the perception of freedom and economic opportunity, though not equitable in reality, while consumerism as diversion reinforces the belief in democracy's superiority.

In nearly every university where political science is taught, a study of Karl Marx's masterpiece *The Communist Manifesto* is mandated. This mandate exists because Marx's work is a treatise for egalitarian government. There is nothing evil about socialism as a political ideology in concept, though it is often maligned as such. The perception of socialism as a failed political ideology persists because of the greed that has perpetuated socialist governments in practice and because of the means by which those governments impose their will on the populace they govern.

China teaches us the power and potential of communism as a political ideology, as it has risen to become a powerful competitor to the United States. Its military and economy are strong rivals to the US, and it is the US that borrows money from China and owes the Chinese government over a trillion dollars. It's not the other way around. Despite China's being an obvious competitor and the threat its military strength represents to the US, the US maintains a robust economic relationship with communist China, despite that relationship resulting in billions of dollars of trade deficit every year in China's favor.

The measure of a civilization's or a government's value is not based solely on its military might or economic prowess. Humans are fundamentally base creatures. It is only through intentionally moral education we rise above those base tendencies. Superficial values dominate the world's populace. Consumerism, fame, beauty, pleasure, and wealth control people's pursuits as people look for fulfillment in things outside of themselves. The potential of humanity will never be fulfilled as a species until we understand that our survival and future prosperity is interdependent on one another. That survival will be accomplished only through a means of living that is sustainable, where humans and the natural world live in harmony with one another. Such a world, where humanity is at liberty to fulfill its greatest potential, will not be in a world divided by arbitrary lines on a map. Squabbles, driven by nations whose disdain for one another is rooted in religious and cultural differences, demonstrates the level of grotesque ignorance our species exemplifies. We are all humans, with a capacity for love and beauty, ingenuity, and creativity that knows no bounds, and yet our world reflects quite the opposite, as brilliant but corrupt individuals have used every possible tool to manipulate and control society and Earth's resources out of greed. We see this same pattern play out in Cuba as it has played out in every government that has ever existed.

Fidel's revolution was born out of inequality. Before Fidel, Cuba was a nation ripe for revolution. Though democratic and capitalist, the island was riddled with tremendous economic and social disparity. In the 1950s, fueled by sugar trade primarily with the US, Cuba ranked fifth in per capita income, eleventh in the world in doctors per capita, and second in per capita ownership of telephones and automobiles. It had a thriving middle class, but it was extremely racially segregated, and enormous poverty among rural Cubans living outside of thriving cities like Antilla, Santiago, and Havana persisted. Sugar cane laborers worked only four months of the year and struggled to maintain survival while living essentially as sharecroppers, perpetually ensnared in cycles of debt. There was widespread illiteracy and malnourishment among rural-living Cubans, and there was a pervasive air of abhorrence among rural Cubans for Cuba's segregated White elite. Indeed, the fields for revolution were ripe.

Cuba's rural inhabitants supported Castro's revolution, though Castro's path toward socialism initially was hidden from them. Conditions were abhorrent for them, and a young, intelligent, and charismatic revolutionary was offering the hope of something better. Rural Cubans followed and supported him. Many of Fidel's solders and trusted allies sincerely wanted to see the lives of impoverished Cubans improved, and some were beloved by the people. Among them were the Marxist Che Guevara, Camilo Cienfuegos, and others who were heroes of the revolution. Camilo Cienfuegos in particular threatened Fidel's popularity among the people, in large part because of his benevolent, jovial nature, which in time made him a potential rival for Cuba's post-revolutionary leadership. His supposed demise in a plane that went missing is thought by many to have been orchestrated at the hands of Fidel.

In time, Fidel's intention to form a socialist government blossomed. In part through the influence of Guevara and the support of the Russian government, Cuba's socialist government was born, though a far cry from the socialism Marx envisioned. In 2016, *Forbes* magazine published an article about the extravagance in which Castro lived, a far cry from the socialism he publicly espoused. Forbes calculated Fidel's net worth at 900 million US dollars ten years prior to the article's publication. While the island lived ensnared in pervasive poverty, Fidel accumulated massive wealth, and his government was just his tool for his exploitation of the people he governed, following the pattern of innumerable dictators throughout history. His government, and its obvious failings, survived through indoctrination, repression, and fear. The US opposition to Cuba was founded on an awareness of his abuses and the perceived threat of communism to US capitalist interests.

The reinvigoration of the Trump administration's travel restrictions and increased economic sanctions against Cuba were in part justified by the administration's opposition to the Maduro regime in Venezuela and the ongoing push for the eradication of communism from the globe. Cuba, to this day, buys oil from Venezuela, so the restrictions on Cuba are claimed as intended to harm Maduro's regime, as well as the ongoing socialist regime in Cuba. Whether it's more likely that America's interest in Venezuela is truly rooted in concern for Maduro's human rights violations or the fact that American corporations are certainly likely to benefit from the privatization of Venezuela's enormous oil and mineral resources, including copious deposits of gold, I'll let the astute reader decide. I'm inclined to think the latter. America is not without its own nefarious foreign policy intentions and is a master at justifying military and foreign policy aggression for the sake of US corporate economic interests. The Iraq War clearly illustrates this.

This Iraq War was portrayed as morally justifiable by the declaration of weapons of mass destruction existing in Iraq. That war led to former Vice President

Dick Cheney's Halliburton making $39.5 billion in US government contracts and $6.5 billion in a contract Haliburton alone was allowed to bid on. Department of Defense contractors received $139 billion in contracts in total from the Iraq War. Whether the declaration of weapons of mass destruction was an egregious mistake made by one of the most, if not the most, technologically advanced and capable intelligence services on the globe, who can say definitively given the nature of how the modern American government functions under the powerful influence of corporate lobbyists.

My work on *The World Hidden* started in Cuba. It was there I had my first interaction with a *santera* and there that I collected the bulk of my firsthand accounts from various individuals, both practitioners and observers of Cuba's shamanic traditions. I count myself fortunate in many ways that my wife's Cuban, and because of that, I have access to parts of the island few outsiders ever see. I have gotten to know the *real* Cuba. Its people, its history, and its political atmosphere have constantly been in my thoughts over the past several years because of my periodic visits there, and my understanding and awareness of life and political realities there has improved because of it.

My wife often jokes that I love Cuba simply because of the crystal-clear turquoise waters and white beaches I've grown to frequent, but for me it's much more than that. The family I've gained, the friends I've made, the unique spiritual experiences I've seen, and the amazing individuals I've been fortunate enough to get to know have all fostered in me the realization of the untapped potential of the jewel that sits just ninety miles off the coast of Miami. What I am still unsure of, however, is what the best path forward is for the nation.

It's not solely my personal relationships and research interests that engender my fondness for Cuba. The island itself represents boundless potential. Whether or not the improvement and acquisition of greater liberties and economic freedoms for the Cuban people will be achieved through the obliteration of socialism or through the willful alteration of Cuba's established government I cannot say. There are great dangers for Cubans regardless of what path their government chooses.

It is the dominant will of Cuban US immigrants to see the destruction of socialism in Cuba. A personal understanding of the way in which freedoms are restricted and poverty abounds on the island cannot be gained fully by an outsider like myself. My wife and countless other Cuban immigrants like her know firsthand the economic and personal freedom restrictions that exist there, in great contrast to America. For many Cuban US immigrants, just the thought of socialism elicits powerfully cathartic feelings of antipathy toward Cuba's government, and particularly Castro. This is one reason why the vast predominance of Cuban US immigrants are Republican, as the Republican Party has historically taken a harder line on Cuba, pushing for greater sanctions and economic restrictions for the nation in hopes of crippling Cuba's economy even further and forcing the government to adopt democracy.

I'm neither a politician nor a political theorist. I know of Fidel Castro and the Cuban Revolution only from books and the older Cubans I've met. I think of the American embargo against Cuba primarily in terms of what it means for me and my friends and family there, and what it means for the research interests on unseen entities I now passionately wish to pursue. Hidden in Cuba is a wealth of ancestral knowledge and a multitude of individuals with extensive expertise of how to interact with unseen entities. It would be a scientific travesty of the highest order to fail to have input from these expert shamans as research is completed in this avenue. The path toward many groundbreaking discoveries may very well rest in their experience and the centuries of knowledge intentionally kept secret. But beyond that, on a more visceral and ethereal level, who among us does not wish to break bread and enjoy the companionship of our friends and family, people whom we love, and pursue the work we are passionate about pursing without obstruction? As I've grown in my knowledge of unseen entities, I've come to realize the enormous scientific value that would come from being able to interact with the many *curanderos*, *santeros*, and others who interact with unseen entities who reside in Cuba.

For me, the chief issue is that America's relationship with Cuba affects my family and friends in many ways. Prior to the Trump administration's increased travel restrictions imposed on Cuba, I could travel to the airport nearest my wife's hometown. The airport in Holguin is about an hour and half's drive from where she grew up. When increased travel restrictions were imposed, we could no longer travel to that airport; instead, all commercial flights were restricted to Havana. That meant instead of a relatively convenient hour-and-half drive from the airport, my family, which includes three small children, would be left to endure a twenty-four-hour bus ride from Havana to Holguin. Anyone with small children can easily understand that such a long bus ride is no small inconvenience, especially since it is only for unnecessary political reasons these circumstances exist.

In addition to the travel restrictions, my wife and I are restricted in the amount of money we can send to our family in Cuba, and we cannot support nonrelatives at all. Cuba is a pervasively impoverished nation, there's no question about that. Still today, the main source of transportation for many in the small town we frequent is horse and buggy. There is public transportation with buses, and there are some who have cars, motorcycles, and scooters, but they are the exceptions. Imposing restrictions on funds for basic things like food, clothing, and transportation for family members seems unnecessarily strict. Even modest amounts of funding can significantly affect the quality of life for Cubans. The Cuban people are a remarkably industrious people, and the entrepreneurial spirit runs deep, as many are already accustomed to finding creative ways to make money to provide for themselves and their family.

Imagine for a moment if the US imposed such stringent restrictions on Israeli US immigrants, restricting the manner of their travel to Israel and their ability to economically support their family members in Israel for

Israel's building of settlements in the West Bank, an action the UN has condemned as having no legal validity and which constitutes a flagrant violation of international law. Such a policy would never exist for Israeli US immigrants, for Jewish Americans possess enormous financial and political influence, and such a policy is obviously unjust. The public outcry would be enormous. The US policy toward Cuba fundamentally represents a human rights violation against American and Cuban citizens, as the aggressive economic sanctions restrict the basic human right to visit family freely and support family financially. If corporations are unlimited in their ability to support political campaigns, as dictated by the Citizens United decision in America, should not Cuban Americans be unlimited in their right to support their own family, many of whom live in abject poverty? This posture subtly intimates corporate rights have reached a degree that exceeds the rights of American citizens, a posture which inherently is concomitant to government-approved corporate imperialism.

Cuba is changing. Fidel Castro is dead. His immediate successor, his brother, is elderly and no longer running the country. The president of Cuba is not a Castro and there is a new generation of Cubans who know about the revolution only through books and the experiences of their elders. There are no nuclear weapons in Cuba. The Cuban people as a whole feel no ill will toward Americans, and when President Obama visited Cuba, many Cubans stood in the street waving American flags, hoping for an improvement in US–Cuba relations.

Things have changed. Cuba has changed. It is time for America to change, too.

Few things foster democracy and freedoms in the way money does. If America truly cared about improving freedoms and democratic principles in Cuba, allowing as much capital into the country as possible and leveraging that economic relationship would be far more effective in improving the lives of the Cuban people as a whole. The American political system teaches us just how much politics is controlled by money. Money buys political influence, and political influence fosters changes in laws. Most often, those changes affect the wealthy, as the wealthy insist on certain freedoms for their own interests, primarily economically. As the lives of Cubans improve through the free flow of capital, inequality would be increased and the desire for economic prosperity among Cuban citizenry would fuel political change even further.

For seventy years America has attempted a stick-only approach to Cuba, imposing every manner of financial restriction possible to manipulate the island's political system toward democracy, and it has failed. Catering to the inherent human desire for prosperity and liberty through the free flow of capital would be much more effective and would improve the lives of Cuban citizens immediately, given proper austerity measures are in place. The benefits would naturally flow predominantly toward Cubans with family outside the country, largely American, who provide material support to them. The imbalance of lifestyle that would develop among Cuban citizens would inherently force the politics on the island to change, as the growing disparity in economic realities would impel the populace toward an economy that more freely allows the flow of capital and a free-market economy that allows Cubans to embrace the entrepreneurial spirit that already exists in great measure on the island.

In the past, Cuba had a deep relationship with the Soviet Union. United by communism and a mutual enemy in the US, Cuba benefitted from trade with the Soviet Union and the Soviet Union benefitted militarily from the geographic proximity of Cuba to the United States. Cuba also sent skilled workers to train Russians in areas in which they were less knowledgeable, such as sugar cane processing. When the Soviet Union dissolved, Cuba went through severe economic depression and hardship, as the country lost its most important trade partner.

The United States still has a growing list of enemies who would benefit from the geographic proximity of Cuba to America. It would be wise for the US to foster an improved relationship with Cuba, if purely for national security reasons. Imagine if China, Russia, or even North Korea fostered a relationship with Cuba for military reasons. The consequences would be as significant as the Cuban Missile Crisis. Instead of Guam or Japan fearing Kim Jong-un's missile tests, places like Miami; Atlanta; South Carolina; Virginia; Washington, DC; and other major US cities and ports would be within the range of an adversarial nation's military. Those national security implications are considerably more severe and substantial than fostering an improved relationship with Cuba, a nation that poses no significant military threat to the US.

America is in desperate need of a foreign policy win. There's an ever-growing list of nations that have suffered at the hands of American aggression, and the world has taken notice. The unjustified Iraq War, labeling poorer nations as *shithole countries*, describing Mexican immigrants as rapists and criminals only helps to fuel the world's hatred for America, and that hatred seems ever growing. In reinvigorating the embargo against Cuba, the US only encourages Cuba to turn to other nations considerably more adversarial toward US interests than Cuba could ever be. In ostracizing Cuba, we invite far more dangerous nations closer to our shores, nations with considerably more capacity to do the United States harm. Desperate nations do desperate things; if China stepped in and offered Cuba financial relief or invested in Cuba's infrastructure for manufacturing purposes or military reasons, the consequences would be enormously deleterious for American interests. A far wiser course of establishing trade with Cuba, providing for their energy needs and other goods the country needs, simultaneously strengthens our national security and weakens the position of other more adversarial nations that might seek to exploit Cuba's weakened economy.

The economic interests of companies that lost land and significant revenue as a result of the Cuban Revolution should be addressed, and their economic compensation could be accommodated in several reasonable ways. A fund that is supported from a tax on travel from the US and other business endeavors, including financial services, could in time produce an appropriate degree of compensation. The Cuban people are highly educated, industrious, and skillful workers. There's no reason Cuba could not be as significant a source for the manufacturing of commercial products as India and China are, with similar or even lower labor costs, and with the added advantage of the geographic proximity that Cuba offers. The labor costs could also be taxed as a source of funding for the economic compensation to losses suffered by American companies from the revolution, and the relationship between Cuba and the United States could move toward normalization in the way America fosters a working relationship with China out of shared economic interests, despite China being a communist country. As a part of establishing renewed economic relations with the island, certain aspects of human rights improvements might be leveraged on behalf of the Cuban people.

Cuba also offers its citizens an effective vaccination for lung cancer, a vaccination the rest of the world would certainly benefit from. While the American embargo endures, US-based pharmaceutical companies lack the opportunity to collaborate with Cuba in developing the vaccination for mass production. Other countries will in time most certainly step in to partner with Cuba and take advantage of America's ignorance. Canada, Spain, France, England, Australia, and other industrialized nations have no such restrictions with Cuba. I frequently interact with tourists from many countries when I travel to Cuba. America is the only advanced nation still clinging to the ignorant notion that Cuba is some dangerous adversarial nation to the world.

I have little economic interest in the pursuit of understanding the world of unseen entities and shamans, yet the potential medical and other technological advances that will come from such study are enormous. There are many areas, including medical technology, product development, military intelligence, archeology, criminal justice, and a host of other fields, that will benefit from the advances that will come as a result of the knowledge gained from understanding these entities. The economic potential is secondary for me personally, but a reality that others more interested in the economic ramifications will likely exploit. The results of such endeavors will likely trump the economic losses experienced as a result of the Cuban Revolution. Fostering a relationship with Cuba for research reasons is essential, for many of the world's most gifted healers and others who interact with unseen entities reside there, and their insights and skills would be instrumental in uncovering the potential benefits and risks that come from these alien life-forms.

There is more than one path for the Cuban people to achieve the liberties and economic improvements desperately needed. I have no political aspirations whatsoever, in Cuba or any other nation including America. I only wish to see the lives of my family and friends on the island improved. The cost to achieve democracy will be a heavy one—the cost of communism persisting, perhaps even greater. But what gives the US, or any outside nation for that matter, the right to decide the fate of Cuba? Democracy has its own significant failings, and the path toward democracy, if driven solely through economic pressure, will quite possibly be a bloody and difficult one.

Should Cuba become democratic under its current course, what would the nation look like after democracy, and during the transition? Revolution is rarely without bloodshed, and in Cuba's case, widespread crippling starvation will likely accompany any civil war, as food shortage is already a problem for the island. Should the revolution become violent, would the US use such violence as a justification for military intervention in Cuba? What then might be the consequences of such US military intervention?

I fear for the Cuban people and the nation in the aftermath of even nonviolent governmental change. Would the rapid shift to democracy and capitalism result in the privatization of Cuba's national resources and beaches? Will a situation develop similar to Jamaica, where private foreigners and corporations own large segments of its beaches and native Jamaicans cannot enjoy the recreational resources and land wealth of their own homeland? Is not such the equivalent of modern-day corporate colonialism and imperialism? Such developments would inherently lead to the same manipulation we see of the US government, where corporate interests and political influence results in the generation of laws and policies in direct contrast to the public's interests in favor of corporations.

Socialism clearly has grotesque failings in its practical existence in Cuba. The Cuban people do not know liberty as Cuba exists today. But Cuba is a fledgling nation: a mere seventy years have passed since its founding. It has really only known one leader in Castro, a man who failed to exemplify the socialist ideals he imposed on the Cuban people. He bled the people dry while leveraging Cuba's resources and his control of its government for his own interests, accumulating massive wealth for himself while his people lived in squalor. Certainly for Cuba's government to follow his example would not achieve the liberty, social, and economic improvements its citizenry deserves.

Seventy years after the US declared its independence, slavery still existed, and Blacks, Native Americans, and women didn't have the right to vote. Certainly such would constitute human rights violations on a scale far more severe than the problems Cubans face today. What America became over time was not a perfect nation but a nation that was allowed to gradually move in a direction that granted more freedoms to and more equity among its citizenry. Though still far from perfect, America's government matured and moved in a direction that provided far more possibility for opportunity for all its citizens, though such opportunity is still not equitable. What will time and opportunity provide for Cuba should it be given the opportunity to chart a new course?

There is an alternative to democracy and socialism that might be feasible and practical for both the Cuban people and its government. Democratic socialism is a relatively modern political theory, where the people choose their leaders, and socialist ideals, particularly regarding health care and education, are embraced. A capitalist free-market economy is incorporated similar to what we see in China. Might such be a better option for

Cuba? It is worth at least considering. A government that embraces and constitutionally protects the rights of its workers would be an achievement even the US cannot claim. A system where Cuba maintains at least partial control over its resources, including it beaches and land, the Cuban people are granted greater freedoms to choose their own course, and liberties like freedom of speech and expression, the right to vote, and a free-market economy that allows for the inherent entrepreneurial spirit of the industrious and educated Cuban people to thrive would be novel. It would be a model for the world, and it is certainly within the grasp of the Cuban people, should they know a governing class committed to the welfare and prosperity of its people.

Time will indeed tell what will become of Cuba. For the sake of science and humanity, I sincerely hope the island will come to know the freedoms and liberties its citizenry deserve, for the world would benefit beyond measure from the wealth of beauty and potential the island and its people represent. Is it for foreigners and others who will face none of the starvation and violence governmental change will likely produce for the Cuban people, many of whom are my family and friends, to express in passionate terms their commitment to democracy and capitalism being established in Cuba? It certainly is their right to do so. And perhaps they are right. But it is easy for a person to insist on changes they will not suffer the consequences of, changes which will produce suffering that neither they nor their family will have to endure. I am not of the mind-set that if the Cuban socialist regime endures, the Cuban people will perpetually suffer the restrictions in freedoms and opportunity that exist. Cuba needs time to become what it could be.

There's no reason to think Cuba's future with socialism is set in stone. There is an air of change in the country, and the seeds of liberty are sprouting. If greater freedoms are won within the context of socialism or communism and the standard of living for Cubans improves economically, would that not be a positive course for Cuba? A government moved with genuine concern for the welfare of its people is the central issue. America has changed considerably since its founding. The restrictions and difficulties in Cuba are nothing in comparison to the human rights violations and subjugation minorities and Native Americans faced in America seventy years into its founding, and the disparities still occurring within American society certainly illustrate it is not a nation without its own skeletons and serious problems.

The progress America has made gradually over hundreds of years was in large part due to the exertion of the populace to force change, not due to the willful granting of freedoms and voting rights initiated by the American government. In fact, in many instances, large segments of American society resisted the extension of the right to vote to women and minorities, for example, and segregation died legally only after great political and social movements. America has its own shortcomings and failings, but the push for other nations to improve the quality of life for their citizenry is expected at a pace even America hasn't lived up to. The Cuban people will achieve greater liberty, one way or another, for the desire for liberty and prosperity cannot be extinguished.

The course of Cuban society will forever be improved if the people are allowed to chart their own course for change. Nonviolent movements overthrew British rule of India under Gandhi's leadership, and Martin Luther King's leadership with nonviolence led to the Voting Rights Bill and desegregation. What might a nonviolent movement in Cuba lead to, even under socialist rule? That movement is starting, and gaining momentum, with the first public demonstrations of dissidents against the Cuban government happening across Cuba in a way it never has before. What will become of such a movement? With the ability of video recording telephones and social media platforms to reveal the abuses of the Cuban government in ways never before possible, nonviolent protest has the potential to be more effective than it has been in any other time in history, given that a vital aspect of nonviolence is its ability to galvanize public support through the demonstration of visible government abuses.

Consider for a moment the social outcry George Floyd's death sparked in America. It led to protests in every major city in the United States, and even in foreign nations including France, Britain, and others, and the Black Lives Matter movement went from one greeted with ridicule and criticism to large segments of the American populace realizing the legitimate concerns of the movement. George Floyd's death was just one of countless murders minorities, and particularly Black men, experienced since the days of slavery and Jim Crow by the hands of the police and White supremacists. What made his death so significant was the gross injustice it demonstrated for the whole world to see because it was captured on video. Cuba's nonviolent movement will be significantly empowered because of the ability of social media platforms to amplify the awareness of the Cuban government's repression of freedom of speech and assembly in a way no previous nonviolent movement has ever enjoyed. The results will be a far more effective nonviolent movement and more rapid change for the welfare of its people.

Democracy in Cuba would certainly produce an increase in liberty and freedom in Cuba. But I fear such change will not be without cost and would certainly result in the same inequality much of the democratically governed capitalist regimes most of the world now knows experiences. In effect, such change could result in Cubans simply trading one system of oppression for a different system of injustice, as they did in the days of Castro. I do not know what path is best for the fledgling nation, nor do I pretend to. I am not an astute enough political theorist to understand the full complexities and nuances of the nation and its needs, or what changes in government might lead to, but I know that change is coming to Cuba. The seeds of liberty have been planted, and their germination may be stifled for a time, but they will not be extinguished. I only hope Cuba's next ruling class will be one born from noble people, individuals committed to the people, and not from a ruling elite whose callousness knows no bounds and whose self-interest will only further exploit the Cuban people.

The potential understanding unseen entities represents is more important than any technological pursuit humanity has every considered, including even AI. What understanding alien life means for our comprehension of the universe is a potential no other technology represents. The understanding these beings represent has implications for virtually every aspect of human life and endeavor, and the existence of political differences that might prevent the pursuit of such study represents a colossal blunder for humanity. America's leadership in this arena will not be enhanced by stifling its relationship with Cuba.

Aside from the national security risks associated with maintaining a hostile relationship with Cuba, Cuba's potential as a manufacturing powerhouse, with its educated and motivated citizenry, also represents outstanding potential for American corporate interests. These considerations illustrate how far the benefits of normalizing diplomatic relations with Cuba outweigh the negatives of maintaining America's ongoing hostile foreign policy approach to Cuba. The world needs the Cuban people's knowledge and expertise in understanding unseen entities, and for the sake of humanity, I hope Cuba's future is one where all are free and welcome to enjoy the marvelous pearl that rests just off America's shores, and the Cuban people are free to pursue whatever course their hearts chart for them.

Author's note: *As a response to the burgeoning freedom movement in Cuba, I wrote the following article, which appeared on my blog www.TheWorldHidden.com on June 28, 2021. With hundreds of thousands of Cubans taking to the streets fighting for their freedom, I could no longer in good conscience support long-term strategies for gaining the Cuban people their freedom over their immediate desire for freedom. A strategy more robust and fast-moving became the obvious only course compassion dictates.*

Chapter Thirteen
Cuba, Biden's Diplomacy Test: Does Elderly Biden Have Any Cojones?

They starved us so much we ate our fear.
— Yomil

Hundreds of thousands of Cubans are now fighting, and many dying, for their freedom, and most Americans don't give a shit. But why? Principally, the media's, and President Biden's, failure to appreciate the historical significance of this moment is responsible. The ramifications of Cuba's freedom movement are enormous for both the world and America, both politically and economically.

At the Free Cuba Fest in October 2020, Gente de Zona took a stand for Cuba's freedom movement. It was a stance some felt was long overdue, given the Cuban-born Latin Grammy–winning group has long understood the realities of the military dictatorship in Cuba. Originally from Havana, Gente de Zona is one of the few reggaeton groups to gain international stardom after starting in one of Cuba's most impoverished communities. It's easy for people to criticize others for not making sacrifices that they individually will not bear any consequence of. Gente de Zona's decision to take a stand against Cuba's socialist regime would have lifelong ramifications not just for them but also for their families, as not only would they likely never be allowed to return to their homeland and family in Cuba, but quite possibly, their family members living on the island would face reprisals for the group's bold stand. Regardless of the group's timing, what would come next for the group reveals the truth that karma has a way of repaying the noble sacrifices we make on behalf of others.

In February 2021, Gente de Zona released the song "*Patria y Vida*," which translates to "Homeland and Life." The song, and its title, would become the rallying cry for the now fervent freedom movement in Cuba that is well underway. For the past two weeks, hundreds of thousands of Cubans all across the island nation have taken to streets chanting "*Patria y Vida*," "*libertad*" [liberty], and "Diaz-Canel *singao*," referring to Diaz-Canel, Cuba's president, with the added modifier of the Cuban word for motherfucker, *singao*. When hundreds of thousands of people line the streets calling the president a motherfucker and crying for freedom, your government has a real problem. I'd seen the handwriting on the wall for Diaz-Canel over a year ago on my last trip to Cuba. Over the years I've made dozens of friends in Cuba, in particular, physicians and artists there, and their dissidence was obvious to me even years ago. But on my last trip, I could feel the change in the air.

Flash forward eighteen months from my last visit to Cuba, and what I felt and wrote would come to the island has come. If people think the protests are going anywhere in Cuba, they've completely misread the room. The handwriting's on the wall for Diaz-Canel, not just regarding his position atop Cuba's government but quite possibly regarding his life. The protests are for freedom, not for food, or for COVID-19 vaccinations, or for anything else the Cuban government can give to make the people's protesting go away. The protesting has recently turned violent, with dissidents attacking police officers, and there's absolutely no reason to think such violence will not worsen should the Cuban government maintain its opposition to the hunger and thirst for freedom nature has endowed into the spirit of every man, woman, and child.

  The internet has fueled the people's appetite for the standard of living most of the Western Hemisphere knows, and no amount of government-sponsored force and violence will be effective at eradicating that appetite. The Cuban freedom movement has heretofore been largely nonviolent, and the world is watching the enormous ability the internet and mobile phones have to amplify nonviolent protests. Nonviolence's aim is to galvanize public support by raising the awareness of atrocities orchestrated by the Cuban government, and that is exactly what has happened, as Cuba's government has responded violently to what are clearly reasonable requests from its citizenry.

  The news media has become predominantly a for-profit industry in America, and as any business with the imperative to increase its bottom line at any cost, what is in the best interest of the public, and journalistic ethics for that matter, now reside in a secondary position. The result is a media system that either swings far to the Left or Right, mirroring the now hyperpartisan American political system, and objective fair news coverage now rests in the hands of smaller independent journalists and online outlets.

Biden is missing an enormous opportunity by responding passively to what is happening in Cuba. America's recent foreign policy strategy is rooted in a series of interventions under the guise of preventing terrorism. Afghanistan. Iraq. North Korea. Syria. American soldiers have died in the thousands in Iraq and Afghanistan under the directive of protecting Americans from terrorism. And, yes, that may be so, for no major foreign-led terrorist plot has met any success in recent years on American soil. But Americans, let's just be honest, aside from those who come from those regions, or have family still there, or business interests there, have little if any real interest in what is happening on the ground in these countries. These countries are halfway around the world, and very few if any Americans travel to these locales in search of a pleasant place to vacation. But Americans do understand freedom. We also like cheaper iPhones, oh yeah, and we definitely love the beach.

Freedom is perhaps the easiest of all PR messages in foreign policy, especially for a nation just ninety miles off America's southeastern border. You have to push hard, public relations wise, to sell American soldiers dying halfway around the world for the abstract concept of protecting Americans from would-be attackers. America's intervention in Cuba, contrastingly, would be something far more easily comprehensible. It is for freedom. What is more American than that? Few foreign policy interventions are ever as easy to pitch as that. Cubans want freedom, and Biden is in a position to grant it. It's that simple. But will he? That is the question.

Should he? Yes, definitely. Politically, he has every reason to intervene in Cuba, but he really just needs to consider only one: Florida. Trump won Florida, largely because Cubans reliably vote Republican predominantly. Republican-leaning Cubans, motivated traditionally by the Republicans' historically harder line on the Cuban embargo, in my opinion, would consider voting for Biden should he aggressively take a stand on Cuba. If Biden were to aggressively intervene in Cuba and develop some sort of coherent strategy for Cuba post-socialism, it would not only strengthen his political position with Floridian Cubans but also demonstrate his skill as both a diplomat and economic strategist.

Democrats don't care about Cuba, but neither do Republicans really, other than Cuban Republicans. Biden would lose nothing politically by being aggressive on Cuba. Democrats are going to support him anyway, no matter what he does. If he's aggressive and, say, cuts off Cuba's leadership militarily, and that leads to the end of socialism, he would win the respect of Cuban Republicans and possibly encourage some of them to vote for him in 2024. Should he combine that with a post-socialism strategy that aggressively supports Cuba economically and encourages and facilitates advantages for Cuban US immigrants, for example, Biden could very well chip away at the Republicans' hold on Cuban

Americans, and that would give him a much better chance of winning Florida.

A part of me thinks that the main reason both the Left-wing media and Biden have avoided commenting on Cuba is because it will invariably lead to Trump getting some credit for the birth of the freedom movement in Cuba, credit he definitely deserves. I didn't care for many of Trump's policies, save for maybe China, but he was right on Cuba. Whether it was motivated by a genuine commitment to freedom or just a desire to give Obama the middle finger by reversing his policies, I can't really say. Whatever his motivations, his reversal of Obama's policies on Cuba directly led to the island's freedom movement. He deserves the credit, regardless of how bad the Democrats and the Left-leaning media don't want to give it to him. What surprises me is how little Fox News has commented on what is an obvious foreign policy win for Trump. I suspect should Cuba actually gain freedom, Fox will be sure to play up the role Trump's policies had on contributing to the country's democratic turn.

Cubans and Cuban American Republicans have a saying for Trump: "Trump *es un pingu*" [Trump has a big dick]. Trump gave it to Cuba hard up the ass with his reversal of Obama's policies. What will Biden do? If for nothing more than his own political aspirations, he should be aggressive in Cuba. His hopes of a second term may very well hinge on a well-conceived, aggressive foreign policy strategy for Cuba. In an election where Republicans threaten to take back both the House and Senate, in part by altering voting laws, Biden taking Florida by chipping away some of Trump's Cuban support may very well be his only real play at keeping the White House.

But what are Biden options with Cuba, and what would something bold and aggressive even look like?

1. Do nothing

Biden could do nothing and simply keep the status quo in place. It would do nothing to help him politically, but it also wouldn't hurt him. Americans have little to no interest in Cuba anyway, but he would also make no headway in chipping away some of Trump's Floridian Cuban support. Biden recently added the caveat of trying to maintain Cuba's internet service and providing medical aid. These are both reasonable and low risk, but Biden was absolutely right to suspect the Cuban government would likely try to make money off such aid, and any change to the embargo would only give a dying regime power to resist what is inevitable. When you have your foot on the neck of your enemy, you don't take it off to give him a chance to catch his breath.

2. Military intervention

Give the protestors guns. Maybe a good idea, maybe a bad idea. The idea is to try to avoid widespread loss of life. Arming the protesters could lead to government change, but it also could lead to civil war, and it's a move that has an outcome that is much more difficult to predict. Recent events in Afghanistan show the unpredictability of armed civil unrest. I suspect Cuban government supporters will not be as motivated to resist widespread civil war and something similar to what happened in Afghanistan could happen, given the armed dissidents made early significant territorial gains. Still, arming another country's citizens looks really bad, both politically and from a human rights perspective, as widespread suffering and starvation would likely affect the island if the civil conflict persists.

3. Occupy Cuba

Really bad idea. Remember Vietnam? The American military has trouble in wars where it's hard to tell our enemies from our friends. Should Russia and China decide to support Cuba, there's no telling what would spiral from that. What began as support for freedom could end up descending into a significant fight in a nation resting just off America's shores. Nobody really wants that.

4. Kill the president and/or higher levels of Cuban leadership

Practical, but politically and morally tenuous. Americans wouldn't care so much, and if is leads to regime change and freedom, an argument can be made that it's morally justifiable. After all, Cuba's president Diaz-Canel is violating his own citizens' human rights. Clearly, assassination is distasteful and morally repugnant, but under the circumstances, it may be the most effective mechanism to effect regime change quickly and result in the least long-term violence to the Cuban populace. Both China and Russia have warned the US about pursing such a course, but given the latest episodes of Chinese and Russian cyberattacks, they've already repeatedly demonstrated their strong commitment to opposing US interests. Little military retaliation would likely occur from such a course, for the US is still *un pingu* itself, both militarily and economically, and much like in the case of Crimea for the Russians, the Chinese and Russians are unlikely to engage in conflict in an area far removed from them geographically and so close to US soil.

5. Diplomacy

"Violence is the last refuge of the incompetent."
—Hardin, *Foundation*, Isaac Asimov

Diplomacy is by far the most practical and morally upright course. If what can be achieved with weapons can also be achieved without them through skilled planning and understanding of the human psyche, certainly that is the better course. A comprehensive guide of changes Cuba must make to remove the embargo should be made clear to the Cuban president and populace. There should be four definitive changes: (1) freedom of speech and the press, (2) new presidential elections, (3) a free-market economy, and (4) freedom of assembly. If the Cuban government meets these four requirements, the embargo should be lifted. Just as essential, Biden should solidify economic investment from the private sector prior to Cuba's meeting the embargo requirements as an added incentive to solidify achieving his demands.

The effort should be international and substantial, as such would galvanize the support of Cuba's citizenry, including current nondissidents, and would maximize internal pressure to accommodate such concessions. The investment would have practical benefits for US residents and businesses who currently use China and India for production of many US-consumed products. The reduction of the cost of shipping and time alone would make an investment in Cuba's infrastructure worth it for many US and international companies with large consumer bases in the US. It would also stabilize Cuba's post-communist economy, ensuring a more peaceful and prosperous nation shortly after government change or diplomatic concessions.

In addition to such demands, the US should take a hard line avoiding and discouraging foreigner-led privatization of Cuba's resources and land ownership. The economic destitution of the nation makes it ripe for corporate and wealthy-investor neocolonialism, and the US taking a hard line against such would demonstrate publicly its interest in Cuba is predominantly one of concern for the nation's citizenry, not the economic interests of gaining the island's resources under the guise of promoting democracy. Cuban citizens should be afforded low-interest microloans that encourage entrepreneurship, which they would most certainly make wise use of, and land should be granted to corporations only with long-term loans of fifty to one hundred years, ensuring foreign investors don't gain indefinite control over the island's vital resources and tourism trade.

A vital aspect of this approach must involve a public address by Biden to the Cuban people. He should call on the Cuban people to demonstrate their desire for freedom on a specific date, asking for their support for the plan he's outlined for their country. The technical logistical challenges of completing such an address must be well considered, and the address advertised for only a brief period, so that the Cuban government cannot plan adequately to contravene country-wide protesting. Coordination with social media leaders of the Cuban freedom movement must take place, which will help to achieve maximum public turnout. With millions of Cubans taking to the streets, with adequate social media coverage displaying the clear will of millions of Cubans in support of Biden by protesting for their freedom, Biden would have all the leverage needed to entertain whatever intervention he desires in Cuba.

Such a strategy would be a novel approach to benevolent foreign policy in the modern era, and a definitive win demonstrating elderly Biden's skill both diplomatically and economically. It would be a foreign policy win the likes of which America hasn't achieved in decades. Perhaps then, the rallying call for Cubans will become "Joe Biden *es un pingu*," and Florida may very well be the prize won for the underappreciated dying art of diplomacy. In an age when inflexibility and callous strength has taken center stage as leadership virtues among Republicans in particular, skilled diplomacy could end up yielding Biden that most-cherished prize of another four years in the presidency.

Excerpt from the upcoming book

ON SPIRITS: THE WORLD HIDDEN, VOLUME II

Interview with a Native American Shaman

As told by Wolf Martinez

WHAT ARE SPIRITS?

Author: *What is a spirit?*

My first thought is what isn't spirit? [laughing] Most people think of spirit as beings, spiritual beings, outside of ourself, and yes that's true. It's the trees, it's the animals, the plants, everyone we meet is spirit, physical and nonphysical. Energy is spirit. The stones have spirit, they have energy, they have a life-force, that's why so many people wear stones. People think of spirit as a place where we come from, and the place where we return to, the source of creation, the Creator, Nature, the Source of Life, the Universe, Great Spirit, God, and so on and so forth.

Sometimes when we pass away, some people when they pass away, and sometimes if it's tragic, they get confused and stuck in this dimension, and their spirit doesn't move on. So there are these things that some people call ghosts or spirits. Some of those spirits are beings that have chosen to stay here. They come here to help. They have chosen to help us as teachers and guides.

There's so many answers to that question, but for me, ultimately, it's more about what spirit isn't. Indigenous People, I would say some Indigenous people, maybe, I don't know, all Indigenous people... I don't know if there's a word *energy* in some of the old languages or *spirit*, those two words were synonymous with each other. And we know everything has energy. Even these things we think of as inanimate objects have energy, there is a vibration.

Second question: What would the relationship between humanity and spirits be like in the best version of human civilization?

Whew— Well, we've lived in that way, and there have been civilizations that have lived in that way. Working in unison, I don't know if working is the right word, living in unison, in understanding that we're surrounded by all these beautiful ones. And calling upon them, honoring our ancestors, honoring spirit, and summoning them for their support and guidance...

Invoking and summoning—if that's the right word, I'm not sure—these helpers, and ancestors, and guides, spirits of the land, for help and guidance; I mean ultimately we do live with them, our language, this language, this patriarchal language has created an illusion of separation. The idea that we don't live in unison with them is an illusion and the language that we speak doesn't open the door for that.

Living in unison with spirit is part of the evolution, the evolutionary process that we are in the midst of, and the direction in which we are going. It is also a part of the place where we've been. I think of the Mayans, Egyptians, a lot of the Native peoples on this continent and the world, the Atlanteans, Numerians, they had a deep understanding of this relationship, and their language reflects that deep relationship.

In our traditions, we're taught to talk to plants, and to animals. We don't just go out and cut the herbs and the medicines without talking to the spirit of that plant and making an offering, without stating our intention or our prayer. Dances are done to call upon the spirit of the deer, the elk, the buffalo before the hunt to honor their spirit, to show our gratitude and reverence, knowing that we cannot live without those plants and those animals, without the sacrifices that they make of their life so that we can live. We honor that life, that spirit. I think that the deepening of a relationship with humanity and spirit opens the doors to infinite possibilities, and if we can imagine it, we can create it. This is also a deepening relationship with our own self, our higher self, our whole self, which includes all that is.

When we are in joy, we are closest to Great Spirit, and we can feel that relationship, that closeness. It is much more accessible when we are in joy. If we meet our pain and discomfort with joy, there's no overwhelm, and there's greater opportunity for learning and growth, it's so much more easily and faster.

Do you have a business where you provide healing for others?

It's so difficult to refer to what I do as a business. It's so difficult to make one's spirituality into a business. You know? I don't feel comfortable describing it in that way. I do private sessions, healing work, counseling, and stuff like that. Some of the work I do is hands on and body work, and amazing things happen. Yes, I do charge a rate for that session. At the same time, I'm really flexible with that rate because I don't think healing or medicine is limited to those who can afford it. That's the problem with our society. I have a mortgage and I have to pay the bills, so it's difficult or challenging. It's an interesting challenge to find that middle road, you know?

I frequently conduct various ceremonies, and yes, people offer donations for ceremony. We don't ever ask for a specific amount. There are no fees for ceremonies. It's just all done by donation and whatever anybody is willing or able to offer as gratitude, as an exchange, for that medicine.

I understand completely. You look at some places like India or Cuba where everybody, for the most part, most of the people are impoverished. You look at what is needed to sustain life, versus America. It's much more expensive. You have to have balance. What you have is a gift and it's valuable and you still have to live in society. You have to balance those needs. Do you mind if I just going to go through some basic background stuff?

Mm-hmm [affirmative].

Where were you raised?

Denver, Colorado.

Okay. And you say you're Lakota. Is Lakota heavy in Denver?

No, I'm Spanish, Navajo, and Ute. I adopted Lakota. The traditions I follow are based on Lakota tradition. You could say I'm Lakota, in a traditional sense.

Okay. How did become exposed to your spiritual tradition?

That's a big question. Okay, my spiritual tradition... My earliest memory is seeing a ghost coming down the hallway toward my bedroom as a little boy, maybe four or five. When I saw that ghost, I got excited, because I thought that Casper was coming to see me. I was so happy to see this spirit coming. It was an apparition, like a whitish fog. But it had form, shape, kind of a human form. As it was moving, I got really happy. Those kinds of experiences were happening almost daily.

How old were you?

Four or five. We lived in that house until I was thirteen. I would say almost daily, I could see them. I could feel them. I was very aware of their presence. I was being raised Catholic, and my mother and my elders were teaching us that there was no such thing as ghosts. I was thinking, "Oh well, they must not know or they must not see." You know? But I was aware that there was something there.

I didn't believe them when they said there's no such thing as ghosts. I was like, "Yes there is." One day they [the spirits] were making a lot of noise and I was trying to watch *Star Trek*. I was about seven years old. I got so upset because they were making so much noise and being distracting. I just wanted to watch *Star Trek*. I was telling them to stop making so much noise and to shut up. In my anger and confusion, I spoke out, "I don't even know who you are, or where you come from, or what you are. My mom says that there's no such thing as ghosts, but I know you're here. Who are you?"

I heard a very clear voice say, "The answers to your questions lie with your Indigenous ancestors." That turned on a light around my wanting to know the spiritual practices of my ancestors, my Native ancestors. My mom didn't know anything about those things. At the age of thirteen, I met my first healer and my first teacher. She was a psychic healer. She was teaching me. She's the first person that said to me, "Oh, the spirits love you and you have a gift."

This was in Denver?

Yeah. She was the first person to tell me that I had a gift. I was so excited. I wanted to learn more about that and how to utilize that gift and how to work with that gift. I started studying about healing, communication, and all kinds of esoteric practices.

What was her tradition?

She didn't have a tradition, per se. She just referred to herself as a psychic healer.

Nobody in your family before her had any experience interacting with these beings?

Exactly. At least nobody that I was in touch with. Looking back, I remember an uncle that worked with herbs and made medicine from herbs. Not just an uncle, but also a grandmother, and a great-grandmother. I'm sure there were other elders, great-uncles, and aunts that did those things, but nobody I was close enough to that had the kind of knowledge or understanding that I was seeking at a young age.

Did she teach you anything? Were you able to make some progress in some way?

Oh, yeah. For sure. She taught me communication with spirits and healing, doing healing work. All kinds of things. Meditation and seeing auras, telekinesis, moving things with your mind. You know, things that were just popular in those years. Telekinesis, I haven't practiced anything like that in so many years, as I don't see the point of it. It's not necessary for healing.

It wasn't until I was twenty-nine that I went to my first Native ceremony. It was a time in my life when I was living in Manhattan and I'd decided to end my life. I'd been doing drugs and alcohol, and I went into that dark, bottomless pit. It felt like there was no escape. It was there, when I was in that place emotionally, that I was invited to my first Native gathering. It was a four-day gathering in upstate New York.

Who was that with?

With a medicine man.

Was he Lakota, or…?

No, he was Mi'kmaq. Mi'kmaq, they're from Canada. He was holding a four-day gathering, and I went to that gathering. I was told to be drug- and alcohol-free for four days before going, and four days after. I hadn't been sober for four days in fifteen years.

Wow.

Yeah. That was a big thing.

You must have felt really compelled to be a part of that ceremony?

Oh, there was no question that I was going to do it. I'd been waiting for an opportunity to go to that kind of ceremony my whole life, since I was a little boy, since that voice said the answers to my questions were with my Indigenous ancestors. That was something that motivated me for many, many years. But by the time that invitation appeared, it wasn't about knowing about the spiritual traditions of my ancestors, it was more about saving my life. Those four days totally changed my life. When I left that gathering, I never did drugs or alcohol again ever.

    I started studying shiatsu, body work, and healing. I worked with a Japanese master. I found another elder that would become my teacher. He was Lakota. That was about a year after I was sober.

What was it about the experience that was transformative for you?

Everything. It felt like being home. I just felt like I had come home. Being a misfit my whole life, home was a feeling I never had. Of course, I felt home with my family. I come from a very, very loving Hispanic family and upbringing. I'm very close with my cousins. But there was always a part of me that didn't feel like I fit in. Maybe that was the part of me that was gay. Maybe that was the part of me that was different. That's why I always felt like a misfit. Those ceremonies, those first ceremonies at that sweat lodge, I really felt like that was where I belonged.

So for the ceremony, he did a sweat lodge? Was that just that part of it?

He did four sweats in four days.

During the sweat lodge, did you interact with some ancestors?

Yes.

Can you remember what that was like the first time? Can you tell me a little bit about what you experienced?

Like I said, it was like coming home. I started seeing spirits again that I hadn't seen since I was young. They welcomed me. They took me. It was like a reunion. It felt like a family reunion with the spirits. I was so happy to be home. When I left that gathering, I didn't have any desire to go back to my old ways. I didn't have any interest in putting myself back into that bottomless pit with drugs, alcohol, and depression.

Did you have an experience, as part of that, that showed you what you should do?

Not yet, not at that point. What I should do? I knew what I wanted to do. It's not a matter of should. I wanted to dedicate my life to these ways. I wanted to learn as much as I could. It had nothing to do with if I should. It was passion.

Yeah. I know exactly what you mean. I've felt that a few different times in my life. It's a hard thing to explain to others. I can't really explain it anyway. It's just something you know you have to do kind of, right? That's what you felt?

Yes.

Okay. Is the Lakota elder the one who showed you how to go perform a sweat lodge?

Well, when I became initiated with him, for him it was just about dedicating my life to these ways. I had found home. I had come home, and I wanted to stay there. It wasn't about wanting to become a ceremonial leader or running sweat lodges. It wasn't about that. It was about participating and doing what felt was my calling. After doing that for about eight years or so, he said it was time for me to start leading the ceremonies. I was like, "Well, I didn't know that I was in this to start leading." But that's how it went. you know.

Is the initiation that you went through something that you can discuss? Did you make any promises as a part of it?

Oh, for sure. I dedicated my life to taking care of the people. I dedicated my life to taking care of these traditions and taking care, as best that I can, of the people and myself, and to bring more beauty into the world, to trust in these ways and these spirits. That's just to name a few. There's a few others, but I don't feel comfortable sharing.

Sure, yeah. I don't want you to share anything that you're uncomfortable sharing. As part of the tradition that you're initiated in, do you promise to not use the gift to hurt people? Is it just for healing?

That's kind of a moot question. Why would anyone? I don't know why anyone would. Of course. I don't hurt anyone. My life is dedicated to helping lift the people, helping to bring medicine and healing, to relieve suffering and pain.

Now, when we talked before, you mentioned that over the years you have interacted with healers from various traditions. I think you mentioned Yoruba and you mentioned santeros… *I can't think of the others you mentioned. I was wondering if you could speak about those experiences, meeting healers from other traditions, how that affected you, maybe what similarities you found?*

Sure. I've had the gift, the huge blessing of meeting many elders and many of whom I'm still involved with, but not all. I've met people of all kinds of traditions, not just Indigenous and Native American, but from all kinds and different walks of life. My focus is on Indigenous traditions. Indigenous means "of the land." My interest has always been in not necessarily Native American traditions, but in Indigenous or shamanic kinds of traditions. I've spent a lot of time with some Yoruba elders. My husband is initiated in those traditions. I've spent a lot of time with people from Shoshone, Lakota,

Apache... Yeah, too many to list... Blackfoot, too many to name, really. All various elders and from different places. I've spent time in Mexico, with elders in Mexico... And Peru, with elders in Peru.

How did you find these people?

Just through life. I wasn't necessarily looking for them. They just came. I met a lot of elders in ceremonies, various ceremonies that I would attend or who happened to be a part of them.

Did you ever feel like your spirit guides were guiding you to interact with different people for certain reasons?

Undoubtedly.

What were some of the reasons you perceived, that you met certain people? To help them in their way? For them to help you in your way?

We that's true of all we meet. I don't believe in coincidence. There's no such thing as coincidence. There's no such thing. It's all part of a greater plan that we're not fully aware of. Some people are drawn together, and energies feel cohesive.

Yeah. Sometimes friendship is unspoken. You just feel it.

Right, exactly. Sometimes? Always.

Did you go through the process of making some sort of sacred space?

Yeah, of course. That's one of the first things you do. You know, you build an altar or create your sacred objects.

Were you guided in that process?

Oh yeah, that's part of the training. My teachers were showing us how to make our altars, our prayer ties, make our *chanupas*, our secret pipes, make all of our medicine things, our medicine shields, the clothes we wear for our ceremony, our regalia. Everything we make ourselves, our drums, our rattles, all of our medicine things.

Of course, when you're young and you're doing it for the first time, you've got to have someone there showing you how to do it. It must be done in a good way, in a prayerful way, in a sacred way, all those things.

Okay. After you started constructing it, what was the consequence? Did you feel like your intentions were supported more, that the things you were hoping to change, you were able to do things more effectively?

Absolutely, yeah. Consequence, that's an interesting choice of word. Everything we do… When we do these things, we're doing it to have an effect. We're doing it for a reason. We don't pray because we don't want anything to change. We pray because we want something to change. Everything we do has a consequence. You can say good or bad, light or shadow. The more conscious we become, the more we become aware that every thought, and every word spoken or unspoken, is going to have an effect. If negative things keep happening in your life, then you might want to ask yourself, "What am I doing that keeps causing these experiences, and what can I do differently to put out energy that is more supportive or joyful?" This is why we must keep looking at our shadows. That is really important. If we're not looking at our shadows, we cannot grow. It's not just our shadows, but our light too.

So many people are so focused on their insecurities, or fears, or their anxieties, and they aren't looking at the beauty that they have. So, they are ruled by those fears, anxieties, and insecurities, and not aware that they are beautiful. If we're putting our energy a lot on our insecurities, the universe is going to reflect all those things back to us because the universe wants us to stop it. The universe wants us to change. The universe is trying to show us ourself.

One of the most beautiful things I've learned or experienced in ceremonies is going into the ceremony and realizing, "Wow, I have a lot of beautiful things. I have generosity, I have laughter, joy, compassion, I have the desire to help... I care." Seeing those things magnified also, which is our light and our beauty. Seeing those things magnified, it's easier to see those shadows and say, "Okay, I see how you've been moving to help me realize who I really am. Not that I am insecure, but that insecurity is actually trying to show me how to be more secure." Meet those insecurities and focus our heart, mind and energy on what we can do to not have those insecurities.

Many of the individuals I've talked with describe having clairvoyance, and it manifests in different ways. Some people have experienced unique dreams. Others can intuit emotions of others and aspects of people's character... What physical ailments a person may face, relationship issues... Without even having any… just being around the person. Is that something you've experienced? Do you mind discussing ways in which you've experienced things like that?

Body language. Intuition… How do you describe intuition? Intuition is a feeling. Intuition is an inner knowing, and it's about learning how to trust that inner knowing but never to make assumptions. It's about having a sense what somebody might be. I'm using insecurity as a metaphor, so I'll just stay with that. Somebody might be feeling insecure in a particular situation, and you can feel that because you have that memory. I have that experience also, so I have a feeling of what this person has gone through and I've been there. But I'm not going to assume that this is what you are feeling or what you are doing.

The way I approach it is that I ask them a question, "Are you feeling insecure right now? Well, this is what's helped me in the past." I don't like to tell people. I don't think people like to be told, "This is what you're feeling. This is what you're doing. This is why you're doing it." It's so invasive. That's not cool. It's more gentle, it's more motherly and nurturing to say, "I have a sense that you're hurting. Tell me what's happening with you." I might know what's going on with them, but I'd rather ask them if they want to share.

Do you feel like your guardians direct you as to the course of helping a person?

Sure. They are the ones doing the work. I'm not actually doing any healing work, I'm just the vessel. I'm mostly doing anything I can to get out of the way so that the medicine can come through. Back to the word you said, clairvoyance. If I get out of the way, then that clairvoyance or intuition from the spirits can come through. They're the ones doing the work.

How does sage assist you? I've burned sage. My wife in her Cuban tradition has used sage in different ways. I know how it makes me feel. Is there a specific reason why Native Americans use sage? How does it assist you?

I actually use cedar more than I use sage, but I use sage too. I use these different medicine plants in different ways. Each one has its own kind of medicine. I use sage daily, but I also use cedar often. Cedar is for cleansing. It's all for cleansing, these two plants we are talking about. We're talking about cleansing. Sage is more physical, cleaning our physical body, our house, our physical space, or our sacred objects. Cedar's a little more for mental and emotional cleansing.

Have you ever used any other assistive herbs, like something similar to peyote or ayahuasca? Are you familiar with these things and their purpose? How would you describe the appropriate use of these things?

Yes, I have experience with some of these teacher plants, as we call them. I've only used them in a traditional setting, in a ceremonial setting.

Not for recreational use?

Oh, when I was a kid. Like I said, I went into that dark, bottomless pit back in those years. Of course I was experimenting. I was addicted to marijuana and alcohol, but I'd also play around with other drugs. It was nothing to do heavy mushrooms, as an example. Mushrooms are a plant, a teacher. In those days, I didn't know that it was a teacher. I just thought it was a way of getting high and hallucinating. I didn't really approach it with the respect and reverence that I've since then learned.

We approach these plants with respect and reverence. They teach us so much. They show us so much. They carry an incredible wisdom and consciousness. They're so intelligent, these plants. That's why Indigenous people, some Indigenous people, have used these teacher plants for many, many, countless generations, including sage and cedar. We use these plants because of their intelligence, because of their wisdom. I feel that the plants are so much more wise than we know. They are so much more intelligent than we are. They have such wisdom that we can learn from.

Like "Take a look at the lily," said Jesus, and he spent much of his time teaching his disciples in the Garden of Gethsemane. Could you tell a difference in the use of the plants in ceremony versus recreational?

Yes. It was more respectful. In ritual it would always show me something or reveal something to me I needed to see.

Do you feel like, over the years, you've become more skilled in your ability to help others? What would you say are ways in which you have grown?

That question can be applied to anybody or anything. You've been cooking for a year or you've been cooking for twenty years. If you've been cooking for twenty years, you're going to definitely have perfected your art. No matter what it is—carpentry, singing, painting, or counseling. Hopefully, you've grown. Whatever it is you're doing, you've grown. If you haven't, maybe go on and do something else if your heart's not in it. By just practicing, we learn. We learn more from practicing than reading. You can't really learn a lot in these ways from just reading. I haven't actually read any books in years.

Are you familiar with, in Cuba, it's described as "the Bad Eye" but in other cultures there are similar names... Are you familiar with this?

No.

Okay, "the Bad Eye" is a bad energy where thoughts of jealousy or people's bad intention toward you can effect general malaise or cause you to have problems.

Who doesn't know that? Everybody knows that. Every time you say "fuck you" to somebody, you're doing that. Every time you say a curse at somebody, you're putting them down. Every time you send a negative thought to somebody, you're doing what you call a "bad eye." Every time you're driving down the road and you get cut off and you curse at that driver, you're sending out bad juju. You're sending out negative energy.

Is it common for Lakota or other Indigenous practices to incorporate water in their sacred space, like a glass of water?

You can't do a sacred ceremony without all the elements: fire, water, wind, and earth. They all have to be present. That's also true with a sweat lodge. We put the stones in the fire. The fire is an essential part of the altar. You heat those stones in the fire. We bring them into the lodge. We put the water. The lodge itself is on the earth. It's on the ground. With the help of the water, we release the breath, right?

Those are all the four elements and the four directions. In the ways that I've been taught in ceremony, in our way and tradition, you have to have all four elements present, which incorporates all the four directions, which is the whole of the universe. All of these things give life. Mother Earth, we couldn't live without her. The wind, the breath of life, we can't live without that. The fire? We can't live without that. The water? We can't live without that. These four simple, basic elements, they are alive. They are beings. They are alive. We honor all of those beings in every ceremony.

During the sweat lodge, is it customary to feed the spirits in some way?

Mm-hmm [affirmative]. Putting up food out on the altar for them. We feed the ancestors almost daily. We feed the ancestors constantly. Also, when we go in the sweat lodge, by singing these sacred songs, we're calling them in. It helps to strengthen them. It strengthens us, by putting our time and energy as a sacrifice. It's an offering by opening our heart and speaking from our heart about our pain and the things we want to get rid of. That also helps feed them, especially by expressing our gratitude. That definitely helps feed the spirits, offering our gratitude.

You know, to answer to your question, what isn't an offering? It's all an offering. When we become aware of that, it's all an offering.

Is animal sacrifice common in many Native American traditions? Like, blood?

I was talking about the intelligence of the plants and the trees. The animals have great intelligence. The eagle doesn't ask how to be an eagle, or what it means to be an eagle. An eagle just is. A tree doesn't ask how to be a tree. The rose doesn't ask how to be a rose. The wolf doesn't ask how to be a wolf. Humans are the only ones that ask how to be human. We're the only ones that have forgotten our original instructions. We've forgotten. Some people remember the original instructions and live by them, which is to take care of each other, and to take care of Mother Earth, the water, and the fire, and the wind. We're caretakers and givers. We're caregivers. We come here to give care because the plants and the animals, they don't need us. They would be much better off without us here, right? But the plants and the animals, they know that in order for us to live, they make a sacrifice of their life so that we can live. Every time you eat, whatever you're eating, whether it's plant or animal, something died so that you can live. Every time you eat, every bite, someone made a sacrifice of that animal's life. That animal made a sacrifice for you.

So, is there animal sacrifice? How would we live if there wasn't? We have to ask the animal to sacrifice their life so that we can live, and the plants. They know that. They're aware of that. That's why every time I have a meal, I offer food to my ancestors, but I'm also making an offering. I always say thank you to the plants and the animals for their life because I know that I couldn't live without them.

When you have students… I imagine you've had apprentices throughout your life, just like you were once an apprentice? What are the most important things you try to impart to your students?

Trust. Wisdom. Trust is so important. We have to trust ourselves. We have to trust Mother Earth. We have to trust life. We have to trust each other. You can't have a healthy relationship without trust, right? That's what life is about. It's about relationships. I feel that's what it's all about anyway. Every ceremony we do, everything we do, is about building trust. It's about building relationship. Relationship with Mother Earth, relationship with the plants and animals, the wind and the fire. The understanding that we're related to the sun, and the moon, and the stars. We're related to all people. Black, red, yellow, or white, we're all related. That's what we say when we finish our prayer. We always say "*Mitakuye Oyasin*," which means "all our relations." When we say that, we understand that we have a place in the web and that everything has its place in the web. When we say "*Mitakuye Oyasin*," we're saying that we all belong, that I belong, and that we're all related.

We all want to be in a healthy relationship. When I use this word *prayer*, it's not about worshiping the sun, or the fire, or the moon, or the ground, or the animals, or the plants. We move to talk to the spirits of the animals and the plants. We move to talk to the fire, or the water, or the wind, or Mother Earth. We move to talk to the stars, the sun, and the moon, to all aspects of nature, because you can't have a relationship without communication. The only way to be in relationship is to communicate.

To have a healthy relationship is to communicate, to open up, to share ourself. *Prayer* is a word that I use but not in the traditional definition of worship. The way I define prayer, it's about communication. It's about wanting to be in communication, in relationship. In healthy relationship, we are open with another. The way I was taught is that we go into ceremony to talk to our ancestors. We go into ceremony to be in healthy relationship.

The elder says, "Hey, say a prayer for us." We do something inside so that our words can come from our heart, from that sacred place. We don't use curse words. In old languages, there were no curse words. People didn't curse each other because there was no concept of that. When we go into the lodge or into the ceremony or whatever, when we'd sit down to talk to our ancestors, our words come from a respectful and reverent way, in a humble way. We have to be humble, we must humble ourself, right? But it doesn't do any good to do that in the ceremony, then go out into life and curse the driver that cuts you off in traffic. It's got to be all the time.

When *you* go home, when *we* go home to meet our creator, to meet Great Spirit, our home on the other side, Great Spirit doesn't ask, "How many times did you go to ceremony? How often did you pray?" Great Spirit says, "How did you live? How was your life?" That's the real ceremony. Every word you speak with your life, that's your prayer. Those are the real prayers. Those are the ones that matter the most. If we're not treating our friends, family, or lovers with love, respect, and care, what's the point? We have to treat ourself with love, respect, and care so that we can treat each other with respect and care.

Part of the reason why I do what I'm doing is because I feel that one of the problems with society is that there's a neglected partnership between humanity and the unseen world. We must help each other. If you don't even acknowledge that there's even an existence of these beings, then how can there be harmony? You see it in the way we treat the Earth, and the way we treat each other. Sometimes I, much like you, feel impelled to do the work I'm doing for some reason. I don't really understand why exactly.

One of the things that I can achieve with my life is to prove, and I know that it can proved scientifically, that spirits exist. It can be done, and water is the key. The reason I say that because when I look at different sacred spaces, particularly with santeros *or* curanderos, *and those are the traditions that I've been most able to talk to with people and observe the most, the water looks different. There is a belief, and I know that it's accurate, that the entities are present in the water in some way. Do you think that if scientific proof is achieved, it would be helpful to humanity? For example, you take this water and you study it, and you prove that these entities are real, would that be significant?*

That's what metaphysics is all about. Metaphysics is a form of science that is trying to validate the things that shamanic people have known forever. I think it's less important for us to prove spirits to others. It's more important for us to prove ourselves to the spirits.

On one level, on a certain level, perhaps level is the wrong word. It's an English word, but it's the only word that I… *layers* is a better word. There's so many layers. There's so many layers and levels perhaps. On one level, I agree with you one hundred percent, and yet, if we are really in deep trust of the whole picture, then we have to trust that everything is for a reason. Think of all the racism coming up, and that has come up in the past few years. Thank you, Donald Trump, for bringing all that up for us to look at it. The beautiful thing about it is that it's showing us how compassionate people really are about racism, and how much people really don't like it and want to change it. It's exposed something, right?

There's a reason behind it. You can hate Donald Trump, or you can hate White supremacy, whatever. Some people hate Black Lives Matter. Hate whatever you want, but that's not going to solve any problems. That's not going to resolve anything. The only thing that's going to bring resolve is to care, to be caring. I remember one day seeing a man that was wearing one of those... it's called a Police Lives flag, a Police Lives Matter flag. I'd never seen one. When I had seen it, I was like what is that? American flag, black and white, or American flag with just blue stripes. I thought, what is that? Somebody said, "It means Police Lives Matter." Does that mean that they're not supportive of Black Lives Matter, or that they're Trump supporters? They said that's basically what that means. My first impulse was to be like, "Oh, I want to go away from that person. I don't want to be near that person. I want to go the opposite direction."

But then when I recognized that, I realized that if I do that then I'll be creating division. My frequent prayer is about creating unity. The medicine wheel of life is black, red, yellow, and white. It creates the wheel. It's the whole. We're all related, all my relations. When I realized that, I realized one of the beautiful things this past year has shown me is the subtle ways in which I've created division and how I judge. If I see somebody that I might think is a White supremacist or a White Republican, wow, how judgmental I am. Maybe they are White supremacists. Maybe they are Trump supporters, or whatever. How can I be creating more unity if I don't interact with this person?

What I've learned is to try and find a way to reach the heart of that person, knowing that we all have more in common with each other than we have different. Maybe our skin is different. Maybe our beliefs are different. Maybe how we have sex is different, but our heart is all the same. We all have the same love, desire, fear. What we have in common is much bigger than what we have different. I challenged myself to go. I want to find the place in myself that can talk to the heart of that person, because that's a person. That's a relative.

When I became aware at a very young age, young on this medicine path, I became aware that there are shamans out there judging shamans. There are medicine people judging medicine people. I was so confused by that. I thought that we are all here to talk to Great Spirit. Our ancestors were all practicing the same tradition. Why are those people at that sweat lodge judging people over at that sweat lodge? Oh, because they think they are doing it the right way, and the others aren't doing it the right way. It's that Christian mentality that has been transferred onto the patriarchal mentality, there's just one way, or the right way, or the best way, and if we're doing it the right way, then you must be doing it the wrong way. That kind of mentality doesn't create unity. It creates division.

I've seen it in the *santero* world, for sure. For sure there's no place I haven't seen it. Santeria, Native American, the Yoga world, all of these yoga teachers are out there saying, "I did it this way and my teacher did it this way." That's Christianity you know, but we don't all have to practice the same spiritual practice. It doesn't matter what you're practicing. If you're a singer or a rock star, maybe you're judging the next rock star because you think you're doing it a better way. But really, the best ones are not in there to be the best. They're in there because they want everyone to win. There's no competition. I want you to win. I want you to thrive. I want you to get it. That's the path of love. There's no competition. The patriarchal way is just warriors fighting.

These traditions that I follow come from matriarchal societies. True warriors train so that they can protect if the enemy's coming or so they can go on the hunt. They train so they can kill a buffalo to feed the tribe. They all train to be the best, but they're not training to compete against one another. They're training so that they can be one band of warriors together, united, strong in mind, body, and spirit. Strong in unity together. There is no competition. Maybe they played games or there was competition, but it wasn't about who could be the best. It wasn't about belittling the loser, but it was about helping everyone to become better.

Maybe this is a human thing. I don't know if it's a human thing. We're talking about spirituality. We're talking about Cuba or New York, wherever. We're talking about these people that are practicing bad juju. What I've learned is that if you're afraid that somebody is sending bad juju, and you're open to being hurt from fear, we have to be very, very careful with that. We have to be very careful with our fear.

Love is the greatest medicine there is. Great spirit, Encantada, God, Buddha, Allah, Jesus, whoever you talk to. All the great teachers symbolize love, right? That's what it is. It's about love. If somebody comes to me and tells me so-and-so is sending bad juju to me, I say, "Oh, wow." I pray for that person. I pray for their heart. I pray for their soul. Good, I'm glad that they're sending it to me, because it isn't going to affect me. I'd rather them send it to me than somebody else, who's going to get caught up in that web. Save some other soul from that bad juju. But who's not sending bad juju in this society? Again, the metaphor of being cut off, bad road rage, that's bad juju. Every time you curse at somebody.

Shamans are people. They're human. They get angry. They get jealous. They also do good work. In the moments that they're doing good work, they're doing good work. In the moments that they're letting their ego, their wounds, or their fears get the best of them, those habits aren't serving them. I try to stay dedicated to growing. You can see the ones that are making mistakes. Pray for them. Forgive them. Love them. But if they keep making the same mistake over and over and over for so many years, they're going to hurt themselves more than they're going to hurt anyone else. Continue praying for them. Continue praying that they wake up. It's not just "they," it's me too. I pray that I'm more awake tomorrow than today. I want to wake up. Who doesn't want to wake up? I want to wake up. I don't want to walk around with my fears, suffering, insecurities, and anxieties. I don't want to walk around with that every single day, all day. That's not going to help anything or myself. We've got to take care of ourselves. The more we do that, the more we take care of ourself, the more we see that it is futile, it is fruitless, that's a better word, it's fruitless to judge anyone for anything.

That's a good way of seeing things, a healthy way. It helps to see things this way.

Are there any prophecies that you get encouragement from? Can you talk a little bit about what encourages you, from a prophecy standpoint?

There's so many prophecies I've heard from so many people, for so many years. For as long as I've been doing these things, for close to thirty years, I've been hearing about prophecies... Lakota prophecies, Zulu prophecies, Yoruba prophecies, Lion prophecies, Shoshoni prophecies, Hopi prophecies, all these different peoples and all their prophecies. The Hindu, Indian, Christian... the list goes on and on because everyone has their prophecies. As I listened to all of these prophecies, obviously, I've spent more of my time invested listening to prophecies of Indigenous people because that is closer to the Earth and feels more grounded. It resonates more with my soul.

But as I listened to all the prophecies, I take the common themes. I take that they're all saying the same thing, right? They're all saying the same thing. Whether you're talking about prophecies or what spirituality is all about, they're all talking about unconditional love. They're all talking about how important our words are and to pay attention to our word and our language. It's so important, spoken or unspoken. These teachings are universal.

As I listened to prophecies, a year ago, this massive spirit came and wrapped around the Earth so rapidly. I was like, "Oh, my God. This is the prophecies coming into manifestation." I recognized it instantly. I see what this is about. This is about rights of passage. This is about initiation on a global scale. This is about change and entering into the next phase of human evolution. This is about all the things that the prophecies have been teaching us all this time. Wow, this is fantastic. I didn't realize I was going to be alive to witness it. "I'm alive not only to witness it, I'm alive," I said. Then the spirits said, "You aren't here to witness this. You've got work to do." We have a part. We have a role. We're here to do something.

Excerpt from the upcoming book

ON SPIRITS: THE WORLD HIDDEN VOLUME II

Interview with a Siddha

Author: An American Yogini describes here initiation with a Siddha Master in a remote mountain cave in India.

I arrived to the cave but he [the Siddha] wasn't there. After a few days he came, and he started the process of an alchemical initiation. He would come and do one part of it with various herbs and plants, and then go away again, and I'd just be in a meditative state for maybe a week, I don't know. Time was hard to keep track of. I had a journal, and I could write in my journal, so I tried to write down the process of the experience. The next week he came, he took all that alchemical substance off of me and then he put another one on, and then that went on for another week or maybe longer.

I'm sorry to interrupt. Did you know why you were going to India, or did you just see him in your dream and he invited you, and you felt that you should go?

Well, that first time he came to me, it was during the eclipse, and I was sick. As soon as I said, "What is your name?" he said, "My name is Hari Baba." I said, "Who are you?", because I don't allow beings to just come to me. I am in charge of my space and who I'm going to talk to or see in my arena, my astral field." So, I said, "Who are you?" again. He said, "I'm Hari Baba."

And then he dissolved into energy and entered into my body and moved around all inside my body, but I knew it was him, though he was just energy. The energy still had his consciousness. I woke up eighty-five percent healed. I was like, "Oh my God. I can move, I can get up, I can walk around. I don't feel exhausted. I'm not in pain." I had been sick and weak for months.

I got in the shower thinking about how long I had been lying down doing nothing, and I was thinking in my mind, "I'm going to make my way back toward working out and getting exercise on a regular basis again," and then he came back again into my vision and he said, "We don't get the enlightened body by working out." He was just always really personable you know.

In meditation I asked one of the other beings [an unseen entity] that I've been connected to for many years if I should go to India, and he said, "Yes, go to him." So, I went. Because I wasn't going to just go.

Wow. You're very brave.

Well, if you're going to be a woman, a Western woman living in India for seventeen years, you have to have the courage of your convictions. I knew why I was going there. I knew what I wanted, what was my goal. It was realization, liberation, enlightened consciousness, and freedom. Absolute freedom has always been very important to me, and the traditions that I studied were all liberating traditions, so you don't have to have all these rules. You don't have to be self-limiting.

We don't have to have all this separation. I had that. I had multiple realizations previously. So, by the time he came to me, I was already teaching other people a variety of things from all that I'd learned, but I hadn't really met any Siddhas. I knew about the Siddhas. Siddha means *perfected*, it means a being who is already liberated. There are Siddha practitioners who maybe haven't attained that yet, but then there are masters of the tradition who are phenomenal in what they can do, in their capabilities.

One interesting thing Hari told me was he had been in that body for a long time. Like hundred and eighty years. He looked very depleted. His body looked very old, and very depleted. He was thin like bones. I don't think he even ate anymore. Some Siddhas come to the point where they don't have to eat, but his eyes were really alive and vibrant, and his energy was really powerful, and there was so much vitality coming through that form.

Ultimately, he started that alchemical initiation process. I felt respectful, like I had to just surrender to whatever was going to happen because I didn't know really what was happening. But I was happening. I'd had other initiations and other powerful transmissions, and I was very intrigued because it was Siddha, and these were Siddha yogis living in these caves, and there were some immortal yogis up there as well. Finally, the whole process culminated on a full moon. Everybody came out of their caves. They did a Homa ceremony and then he gave me my name. They were all chanting my name and making offerings into the fire to connect me to the lineage, and then they chanted the names of all the people in the lineage including Hari.

At the end of that ceremony, I realized that this master was going to leave that body. They have ways of moving in the world. They have a tradition of immortality. There're several ways that's achieved. First of all, you have to have longevity. You have to stay alive long enough to be able to learn to master all the processes. So, they do a lot of things towards longevity. Practices, internal cleanses, herbs and dietary things, and cave retreats and isolation, things like that.

They can reverse age by twenty years. They look twenty years younger. Some of the beings are still regenerating in the same body, they might be eighty-five years old but after they do their cave practice and they come out, teeth that have fallen out are back. Wrinkles are gone. They're healthy and vital again. There's all these technologies that will keep you in the body, but the next level of immortal understanding is to understand your consciousness is fully free and it can move out of your body.

Most mystics and practitioners have experience of going out of their body for different reasons. So, they do what's called transmigration. If the body that they're in becomes too old and they really can't regenerate it anymore, they do the technique called transmigration where they typically will pick another yogic practitioner who maybe passed from the body in a cave somewhere and they'll inhabit that body and merge it with their own consciousness and make it come back into life again. But it's their same consciousness, their same being, and then that body will go through the kaya kalpa process of regeneration to bring it into their level of consciousness.

Hari was a transmigrator. He had been in that body a hundred and eighty years, but he had told me he was going to transmigrate into another body. I think maybe he had actually left that body before I came, because the condition was very frail. I think he reanimated that body. He reentered that body just to give me the initiation. I asked him, "Why are you initiating me? You have all these disciples around you who are Indian people, yogis that have been up here a long time." He said, "The Siddhas want a woman disciple. A Western woman disciple. So, you have all the background for that. You studied all these years and practiced all these years." So, I became his disciple through that initiation.

I want to just let you know, first of all, when I was a child I already had mystical vision. I was already able to see the invisible beings.

How old were you when you had your first experience?

Since I was born I remember things, since I was born. My first real close contact experience was when I was four years old. I made friends with this little girl from down the street. Her name was Margaret. I can still see her to this day. She always wore a little cowgirl dress-up outfit.

She would come over and we would just hang out. She lived in a house down the street, and my mother would say to me, "Who are you talking to?" And I would say, "It's my friend Margaret." And I couldn't understand why no one else could experience her. She said, "Well, where did Margaret come from?" I said, "She lives in that house." Well, nobody actually lived in that house, that house was empty.

My ancestry is very interesting. My mother's people were Celtic and Pict, from the Irish and Scottish clans. Ultimately, over generations they married into the British people, and they were suppressed. Their traditions were suppressed. But more poignantly, my father's people were from the South of France. They were from the Languedoc, and they followed the mystical traditions of Egypt, and the mystical traditions of the original teachings of Jesus and his followers. They were the followers of Jesus who all migrated across the Mediterranean from the Holy Land. There are Isis ruins in the South of France.

So, there's a lot of Egyptian influence in my blood. My European ancestors were mystical people, and when the Inquisition was organized by the Catholic Church of Rome, first the state of Rome, and then the Spanish because they were very Catholic, they killed millions of people over six hundred years. They killed nine million people. Most of them were mystics, women, herbalists, and healers. They just murdered anybody who was a heretic in their view. These parasitic European people were under the influence of that Roman imperative and ended up infecting the whole society of Europe. They killed all of their own Indigenous tribal people and mystical practitioners without mercy, burning them to death. And then they went out in the world to conquer all the other Indigenous people.

When I hear people talk about that issue I think, 'They don't even know their own history. They don't even realize they killed their own mystics and healers and shamans, first in their homeland, and then they went after all the rest." So, you could be a White European person, but your bloodline could lead to the mystical people who were murdered. And that's my father's family from the South of France. Since that time, I've made a lot of pilgrimages to the South of France.

My father's last name is very common over there, and I followed the Templar Trail, and I followed the Trail of the Cathars, and all the mystical holy people who were killed en masse. I stood on top of Montségur in the South of France, which is a mountaintop that was a stronghold of the Cathars. The Cathari were peaceful, sweet, nonviolent, and they didn't eat animals because they were vegan. They had a holy tradition as a most elevated, realized people.

They were educators and would educate not only their own people but the children of the Romans, and everybody else, because they were intelligent and benevolent people. I stood up there where they had taken the last four hundred Cathars and burned them all to death in the central plaza, one after another. It is said some of the Cathari women, the Adepti, ran to the fire themselves and threw themselves on the fire rather than be led by their enemies to show them, "I do not fear death. You think you're going to stop this energy by killing us, but we know better, so we freely burn ourselves."

They also had the power of self-immolation, so they could set themselves on fire if they needed to. And some of them did that to protect the Cathari who were running away with the gospels, the manuscripts, which is what they were after.

I was born in a Catholic family, just like all my ancestors had been forced to become Catholic or die. That was our only option. That's really how the Catholic religion spread around the world; it was by them just killing people. It's our religion or the sword. And in the Gospels, they flat out identify the Pope as the Antichrist and the Catholic Church as the Antichrist. It's very suspicious, especially in the Gnostic Gospels that have come out more recently.

So, I grew up being Catholic, but my mother's family had many visionary women in their lineage, and one of her aunts was forced into an insane asylum because she had mystical visions and they thought she was insane.

That still happens to this day. Seeing what happened to her aunt terrified my mother and made her afraid of her own spiritual abilities, so when she saw that I had a friend who was obviously a girl who had died in that house, she was scared. She wasn't scared of what was happening with me, she was scared of what would happen if people knew the gift I had. Some of her ancestors were killed for just having the gift of sight. She kept an eye on me, but she didn't know anything about how to guide me or teach me how to use what I had. I was just a very small child, and I thought everyone had that experience. I didn't understand that others couldn't see spirits, and people have the tendency to project that onto you. "Oh, you're just imagining things. It's an imaginary friend." My mother took comfort in the Dr. Spock books, that these are just imaginary friends.

We ended up moving from that neighborhood to another neighborhood a few years later, and my mother said to me one day after we'd been there several months, "Whatever happened to Margaret? You never talk about Margaret anymore." I said, "Margaret lived in that house. She lives there. She doesn't live here." So, that kind of blew up that child psychology imaginary friend theory.

After we moved, my best friend in school was a seven-year-old girl who had been diagnosed with brain cancer. It had stunted her growth, so she was really small. She'd lost all her hair from the treatment, so she wore wigs. Nobody wanted to be her friend. My childhood playmates were angels. I would experience angels around me constantly, and I would play outside by myself and be swinging, and the angels would be laughing. I would be having a great time, and my mother was like, "That's so strange?! Why is she so happy playing by herself?"

When I went to school, it was traumatic. I was sick for days. I couldn't tolerate the environment; the environment was so toxic. I just wanted to go home and be outside with the angels, but I was forced to stay in school. When I met Sandy in the second grade, she was an angel. I could tell. This is an angel in a body, and I became like this with her [crossing her fingers], and she was my best friend for two years. When we were in fourth grade she died. Before I knew she'd died, I woke up early one morning, and there was Sandy with beautiful natural hair. She looked amazing. She was with all the angels.

I was so happy because I'd watched her suffer from the treatments and everything else, and sometimes she didn't have any energy. So, I saw her like that, and I was so happy and excited, and I ran out to tell my mother. My mother was hanging up the phone from Sandy's mother telling her that she'd passed, and my mother was trying to figure out, "How am I going to tell her that her only friend, her best friend has passed away?

She said, "Sit down, I need to tell you something." So I sat down, and she said, "You know Sandy was really sick, right?" I said, "Yeah, yeah, yeah, but let me tell you this." And she said, "Just let me finish what I'm saying." So, she tells me, "Well, you know she was so sick she couldn't stay in her body anymore and she went to Heaven." And I said, "I know. I saw her this morning. I saw her with the angels, and she has hair and everything!"

I was in that childlike joy, and my mother was like, "Oh, my God!" And she called Sandy's mother back and she said, "My daughter's telling me something that I think you should hear." So, my mother saw that my gift could be compassionate towards a mother that had lost her child, so she took me over there, because Sandy's mother wanted to see me. When we got there, her mother asked, "What do you want to tell me?" So I told her. It brought some comfort to her. It was an amazing experience, it really encouraged her, and it was really good for me that my mother did that. Sandy's mom was really grateful, and it was very comforting to her to know that somebody could see her and especially since it was coming from a child and she knew there was no agenda. But my mom still wasn't hundred percent comfortable around me at that point. After that, I never had grief when someone passed, because I knew there was liberation after death.

Wow.

I had no idea why I had that ability and nobody else did, or why I saw things other people didn't. But I got used to it after a while, and I stopped talking about it to others. I learned to be quiet about it, but it never went away. When I was thirteen and we were in Mass one day, I had another pivotal experience. Between ages thirteen to fifteen is the time kundalini first rises in everybody. At the first onset of adolescence, there's a release of hormones, and sexual energy activates, that's the first time kundalini comes alive. That's why it's so important for teenagers to have a spiritual resource where they can learn about these things, and not just some church that's telling them sex is a sin and all of that. They need to understand that it's their spiritual energy. That's what kundalini is. It can cause them to have spiritual experiences.

So, I was in church, and Mass was happening, and all of a sudden beside the statue of Mother Mary, these three saintly spiritual beings emanated out from that statue. It was Mother Mary, Saint Elizabeth, and Saint Margaret. There they were, alive. They were all full of light and they were coming right at me. They all came to me, and they surrounded me, and I went into an ecstatic trance. I was moving and moaning, and the priest was looking at me, and my mother was like, "Oh my God, now what? What is she doing now?" But I was having a true mystical experience with these powerful spiritual beings.

After that Mass my parents had a meeting, and they sat me down and said, "We're going to send you to this Catholic school." All my sisters were in public school, and I said, "Why are you sending me to Catholic school? I'm not doing any of the bad things they do." From the time I was a kid, I had a strong moral guide, and I wouldn't do things my sisters would do, and they didn't like me because of it. They would pick on me and they would call me, "Oh, she's a perfect little angel. She doesn't want to do anything." And I would be mocked for it, but I just knew I'm not going to do anything that's not right. So, I thought, "Why are you putting me in Catholic school? I'm not the one doing bad things."

But it turned out it was a boarding school with Franciscan monks, run by the Franciscan order who are mystics. And they all had beards and long hair, and it was the time of the hippies, and we were coming into the transition between the late sixties and early seventies. It was great. It ended up being the best thing she could have ever done. Because they knew. They understood all of it.

They taught us about the saints and their mystical experiences. They taught us about states of ecstasy and altered states of consciousness. They were so liberal, and so progressive, and so spiritual, and that way I didn't lose my gifts. I also didn't get into it with the Dominican order, or the Jesuits or something, which were running the parishes. So, she hid me in plain sight sort of. It protected my gifts. It was a real blessing in the long run. But of course, when I was thirteen, I didn't think it was such a great thing, but in the long run it was. I had many powerful spiritual experiences during my time there at that school, so my mother was really watching out for me even though she didn't really understand what was going on.

When I was in my younger twenties, I got married and had children with a man who was Indigenous. He was half Mestizo-Castellan Spanish from his grandmother's side, and he was half Karankawa-Apache and Mayan from his mother's side. My dad, a Frenchman, knew about everything what happened to Indigenous people in this continent, and he was adamant that we were going to be educated about it. It was the worst atrocity that ever happened in his mind, and he was furious. He was so passionate about it. Every single weekend my dad would take us to all these different tribal reservations in Arizona and he would say, "If you want to buy jewelry, or artwork, or anything like that, you come here and you buy it from these people. You don't buy it from those White people that are running the stores in town." So, he would introduce us to different Indigenous people he knew, and take us to their ceremonies and he'd say, "Look what they've done to these people. This is the worst thing that ever happened."

He was a spiritual person, sounds like.

He was a spiritual person. He was conservative outwardly because it's hard. So many of his ancestors had been killed for that. You know? He was just a kind-hearted person, and he was very sensitive. He imprinted that on us, that these Indigenous people never deserved anything that happened to them, and they're still being treated badly. We were to never lose awareness of that our whole lives, and we were taught to honor them, and respect them, and spend time learning from them. He was powerful in that passion.

My husband's people weren't necessarily warrior-like. They weren't out to try to take land or anything, but they would definitely fight, especially to defend themselves. The Apaches are very fierce, and my husband was very fierce. My children are fully one quarter Karankawa Apache and their DNA shows Mayan blood also. I ended up marrying an Indigenous man, maybe because my father taught me that these are people who are good.

Did he (your husband) have a spiritual practice?

He was a very spiritual man. His mother was a *curandera* and his father was a brujo, so he was a mix of light and shadow. We just had a very natural lifestyle. We had some land and raised the children virtually outdoors all the time, and he taught me so much about how to be a natural parent like his mother had been. She had eleven children.

When our children were small, he died. That was my first big life initiation, to only know him through the spirit realm from there on. It was devastating, so difficult. I was young, and my children were very young. It was a huge event, a huge loss. We moved to the city, and I went back to school and finished getting my degrees. I studied clinical social work because I wanted to help people with psychological and emotional pain, but I also wanted to understand the whole depth of human psychology. I had already, of course, a spiritual concept towards life, and when I was able to open my own clinic, I totally devoted my practice to treating trauma and addiction. I eventually learned about intergenerational trauma, and that made even more sense why the tribal people today still carry genetic trauma in their DNA.

No one really addresses that. I worked with everything from Holocaust survivors to torture survivors, combat veterans, victims of crime, parents of murdered children, and victims and family members of victims of mass shootings. I had a critical incident response team that would go and respond to those incidents. But in my practice, spiritual things would happen all the time in my private room where I had the survivor with me, where I was counseling them one on one. I'd learned early on there's nothing you can do. There's nothing you can do if a human being is harmed by another person and they're suffering pain from it, or they've been exposed to death and destruction and their soul is wounded.

I took some training from native shamans too, about the wheel of life, and how they believe when a trauma happens the soul gets thrown off the wheel and the shaman has to go and get the soul and retrieve it. But I was also starting to study the Western mystical traditions, which are kabbalah, alchemy, ceremonial magic and ritual, and the Christ figure. It doesn't have to be Jesus, it could be Apollo, it could be any being that dies and is resurrected. That kind of mythos. Osiris is a part of that.

I also studied theurgy. Theurgy is where we assume the god in our bodies, like the Egyptians did. So, I was really coming back to my ancestors' teachings without knowing that at the time. In those early days of my practice, I was studying in a correspondence school, a mystery school, and I was doing my clinical practice at the same time, so I was able to step back from thinking exclusively like a clinical person who has to have all the answers. I never took notes.

It's offensive to write notes in front of a person whose heart is bleeding. Only the survivor is the true expert of their pain, and only their own soul can heal them. So, I would do dramatic things that. I would take them out in the woods at night to ceremonially bury the pain. Symbolically reenacting something that's powerful on the subconscious mind helps understand that the shadow is just a veil. It isn't about what it appears to be on the surface. It serves as an initiation to lead you into a higher understanding, a higher point of view. Sometimes I would even sit on the floor at the feet of my client to give them their power back, to make them be the one with the power. Some of them were quite terrified of that. They would cry and beg me to get up, but I would say, "No, it's okay. You're going to just relax. It's okay for you to have the power." I did a lot of controversial kinds of healing, and brought in body workers to release the trauma from their body. I was way ahead of my time back then.

It was a very powerful spiritual experience working with those survivors, for them and for me, and I became famous because I had such a high success rate and people were flying in from other states to come and see me in my clinic. These professional organizations and universities were contacting me and saying, "We want you to come and train these MDs and train these PhDs." I had to really be on top of my game clinically and get comfortable speaking to intellectuals. I was able to teach them a few alternative things, like how important art therapy can be, and how important it is not to disempower our patient. We have to be careful not to replay the trauma dynamic.

Eventually, I was on television, working with the US Senate, and writing papers on how trauma and addiction treatment should be done. I was really one of the beginning movers and shakers in the whole arena of post-traumatic stress. It was a brand-new diagnosis at that time. I was able to help a lot of people by helping change those systems, and I was spiritually evolving too. My practice was going really well, it was very successful. Often, angelic beings would come into my sessions, and I was just in awe of it. How powerful the spiritual realm is!

One day I had a very intense session where an archangel had come into the room. The client was affected by it, she experienced a massive healing, and we were both kind of in an altered state. She was going out to the reception desk, and I came out to the hallway and I looked toward her. She was paying her copayment at the desk and something turned inside of me, and I thought, "Why am I getting paid for this? What am I doing? How can I be charging them? I didn't even do anything. Why is she paying my office?"

I became very disturbed by that experience. I couldn't reconcile it. I mean, yes, I was renting a space, I was holding an energy, I was bringing something other people weren't bringing, but it still felt off kilter. So, I went home and designed a ritual. I was just a novice practitioner at that time, and I didn't totally know what I was doing. I didn't even tell my mentor that I was going to do it. I just decided I wasn't going to work for money anymore. I told myself, "I will work, I will share my gifts, I will do what I came to do, but I will not have in mind that I'm going to have to make money." Mystical teachings were elevating my mind, they were expanding my consciousness so vastly that I said to myself, "There's so much money in the world, there's an endless, amazing, huge amount of money in the world. What makes me think as a little ego, I can't do something for it, and I'm not going to get it? If it's out there, I can make it come to me.

"That's it," I thought. "That's what I'm going to do. I'm going to do an active magical ritual with my intention, and I will no longer have to work for money. Money comes to me." But I didn't know enough to put some protection in there, so I did the ritual, and it was powerful. But nothing really happened until a year later.

I closed everything down, and sat at my altar. I prayed, meditated, and worked every single mystical practice I'd learned over the years, but had been too busy with the trauma center to practice. Soon, my lawyer told me, "You'll never have to work for money again." I was like, "Oh, my God. It's the ritual. I did it to myself." This is the power of ritual. That was it. I didn't have to work for money anymore, and look at how many doors that opened. Later I learned there are ways to do that without having to go through trauma to get there. You can say, "This will happen without causing harm to me or anyone else." That would have been a good thing to add in there. But then my mentor said the universe doesn't have the same definition of harm that you do.

The play and the Maya duality is just going to happen the way it happens. A legal situation was the dweller on the threshold, and I walked through the doorway. There are always moments like that, but they're different ways of going about getting what you want. Anyway, that was my spectacular exit from the working world.

I spent five years doing only spiritual practice after that. I did more rituals in my circle. I learned much more about magic. I learned more about Kabbalah, and who we really are. What is the psychology of the human? In truth, this body is the least of it. All these emotions, all these mental beliefs, are on the lower level. But the true self, the true genius of Christlike power and kabbalah gives us a way to attain that. It's a very powerful system of spiritual psychology.

And I liberated myself and I had experiences of self-realization where my whole aura would be lit up with vibration and energy and power and that five-year retreat was really important to my growth. It led to me meeting the man from India.

Wow. You took a bad experience, and you turned it into an opportunity to be of service. A lot of times something happens, and opportunity and financial freedom comes and then people use it for something other than service. That's beautiful to me.

After five years in meditative retreat, practicing esoteric tradition, then I went to India and stayed there for years.

What did your kids think? About you going to India?

They ended up being raised like this because I was already with an alternate spiritual awareness. From the time they were born, I could see who they were, so I raised them to be who they were and fully support their powers. My daughter also had vision when she was born. My son also could see. And then, when their father passed, they had more experiences with their ancestors, and he would come to them often. So, I took them to make altars and I taught them all the things I was learning. I taught them tarot. All their friends came to my house, all these teenagers. They were fascinated by all of it, and they were free to learn and observe. They could explore and they could experiment and one time I said to them, "Why don't all of you get together and make a ritual for the summer solstice, and you can perform it in the stone circle tonight?"

Oh they loved it, they all got together. The girls dyed some sheets different colors, and it was very elaborate. It was all about the Sun God. They orchestrated the whole thing, and the girls were all together, and the boys all were together, and this one guy was the Sun God, and they painted his face, and they painted the other boys' faces. The girls painted their faces and put crescent moons on their forehead like Celtic priestesses. They just did this amazing, elaborate, nature-based ritual, and that was kinda how my kids all grew up.

I used to go back and forth between New Zealand and India. I had a place in New Zealand because you could only be in India six months at a time and then you have to go out for a few weeks before you could get another visa to come back. So, I had New Zealand as a place because I couldn't come back to the US. It was too much energy. But in New Zealand I lived close to the beach, and it was quiet, and I could continue my practice.

That's the challenge of most spiritual paths. If you're going to take it all the way, which I had no idea what that even meant at the beginning, you have to get passionate about it. It's your calling, and it's just like what Jesus taught in the Gospel, you have to leave everything. Leave your family. You have to leave everything for enlightenment. You know, one of my mentors in Mumbai said to me, "The cost of enlightenment is nothing less than everything."

And it's true. You don't live the same, you don't behave the same, you don't sit anywhere the same, not that I ever did. If you decide to go all the way, you go all the way. Then all of humanity is your family. Not that I don't still see my children and my grandchildren. I do see them. It's good for children to grow up knowing the power of a ceremony or a ritual is in your own hands. You don't have to go to church and watch the priest do it.

Thank you for reading *The Illusion of Superiority*. If you enjoyed the book, I ask that you please leave a review online so that others might be encouraged to share your experience.

I invite you to visit my official website

www.TheWorldHidden.com

to view my photography, blog, and excerpts of my other literary works.

www.ingramcontent.com/pod-product-compliance
Lightning Source LLC
Chambersburg PA
CBHW071348210526
45465CB00001B/14